Springer Textbooks in Earth Sciences, Geography and Environment

The Springer Textbooks series publishes a broad portfolio of textbooks on Earth Sciences, Geography and Environmental Science. Springer textbooks provide comprehensive introductions as well as in-depth knowledge for advanced studies. A clear, reader-friendly layout and features such as end-of-chapter summaries, work examples, exercises, and glossaries help the reader to access the subject. Springer textbooks are essential for students, researchers and applied scientists.

Maurizio Petrelli

Machine Learning for Earth Sciences

Using Python to Solve Geological Problems

 Springer

Maurizio Petrelli
Department of Physics and Geology
University of Perugia
Perugia, Italy

ISSN 2510-1307   ISSN 2510-1315   (electronic)
Springer Textbooks in Earth Sciences, Geography and Environment
ISBN 978-3-031-35113-6   ISBN 978-3-031-35114-3   (eBook)
https://doi.org/10.1007/978-3-031-35114-3

This work was supported by University of Perugia

Cover illustration: ipopba / stock.adobe.com

This Springer imprint is published by the registered company Springer Nature Switzerland AG
The registered company address is: Gewerbestrasse 11, 6330 Cham, Switzerland

Paper in this product is recyclable.

To my family and friends

Preface

Machine Learning for the Earth Sciences provides Earth Scientists with a progressive partway from zero to machine learning, with examples in Python aimed at the solution of geological problems. This book is devoted to Earth Scientists, at any level, from students to academics and professionals who would like to be introduced to machine learning. Basic knowledge of Python programming is necessary to fully benefit from this book. If you are a complete novice to Python, I suggest you start with Python introductory books such as *Introduction to Python in Earth Science Data Analysis.*[1] *Machine Learning for the Earth Sciences* is divided into five parts and attempts to be geologist-friendly. Machine learning mathematics is gently provided and technical parts are limited to the essentials. Part I introduces the basics of machine learning with a geologist-friendly language. It starts by introducing definitions, terminology, and fundamental concepts (e.g., the types of learning paradigms). It then shows how to set up a Python environment for machine learning applications and finally describes the typical machine learning workflow. Parts II and III are about unsupervised and supervised learning, respectively. They start by describing some widely used algorithms and then provide examples of applications to Earth Sciences such as the clustering and dimensionality reduction in petro-volcanological applications, the clustering of multi-spectral data, classification of well-log data facies, and machine learning regression in petrology. Part IV deals with the scaling of machine learning models. When your PC starts suffering from the dimension of the data set or the complexity of the model, you need scaling! Finally, Part V introduces deep learning. It starts by describing the PyTorch library and provides an example application for Earth Sciences. If you are working in Earth Science and would like to start exploiting the power of machine learning in your projects, this is the right place for you.

Assisi, Italy   Maurizio Petrelli
28 July, 2023

---

[1] https://bit.ly/python-mp.

Acknowledgments

I would like to acknowledge all the people who encouraged me when I decided to begin this new challenging adventure, arriving just after the satisfying but extremely strenuous challenge that was the book *Introduction to Python in Earth Science Data Analysis: From Descriptive Statistics to Machine Learning*. First, I would like to thank my colleagues in the Department of Physics and Geology at the University of Perugia. I would also like to thank the Erasmus Plus (E+) program that supported my new foreign teaching excursions in Hungary, Azores, and Germany. Namely, Professor Francois Holtz (Leibniz Universität Hannover), José Manuel Pacheco (Universidade dos Açores), and Professor Szabolcs Harangi (Eötvös University Budapest) are also kindly acknowledged for allowing me to teach the "Introduction to Machine Learning" courses at their institutions. In addition, I thank J. ZhangZhou (Zhejiang University) and Kunfeng Qiu (China University of Geosciences) who invited me to give lectures and short courses on topics related to the application of machine learning to Earth Sciences. I would like to acknowledge the "Piano delle azioni collaborative e trasversali" at the University of Perugia with emphasis on the working packages "3.1 - Disastri e crisi complesse", "4.1 - IA Data management e Data Science", and "4.4 - Scienza dell'Informazione e Calcolo ad alta prestazione." Professor Giampiero Poli is kindly acknowledged, thanks for being a great mentor during my early career. Finally, I give my heartfelt thanks to my family, who, once more, put up with me as I completed this book.

Overview

Let Me Introduce Myself

Hi and welcome, my name is Maurizio Petrelli and I currently work at the Department of Physics and Geology, University of Perugia (UniPg) in Italy. My research focuses on the petrological characterization of volcanoes with an emphasis on the dynamics and timescales of pre-eruptive events. For this work, I combine classical and unconventional techniques. Since 2002, I've worked intensely in the laboratory, mainly focusing on the development of UniPg's facility for Laser Ablation Inductively Coupled Plasma Mass Spectrometry (LA-ICP-MS). In February 2006, I obtained my Ph.D. degree with a thesis entitled "Nonlinear Dynamics in Magma Interaction Processes and Their Implications on Magma Hybridization." In September 2021, I authored the book titled *Introduction to Python in Earth Science Data Analysis: From Descriptive Statistics to Machine Learning* published by Springer Nature. Since December 2021, I have been an Associate Professor at the Department of Physics and Geology at UniPg, and I am now developing a new line of research for applying machine learning techniques in Geology.

Styling Conventions

I use conventions throughout this book to identify different types of information. For example, Python statements, commands, and variables used within the main body of the text are set in italics. A block of Python code is highlighted as follows:

```
1  import numpy as np
2
3  def sum(a,b):
4      return a + b
5
6  c = sum(3,4)
```

Shared Code

All code presented in this book is tested on the Anaconda Individual Edition ver. 2023.03 (Python 3.10.9) and is available at my GitHub repository (⟲ petrelli-m):

     ⚭ http://bit.ly/ml_earth_sciences

Involvement and Collaborations

I am always open to new collaborations worldwide. Feel free to contact me by email to discuss new ideas or propose a collaboration. You can also reach me through my personal website or by Twitter. I love sharing the content of this book in short courses everywhere. If you are interested, please contact me to organize a visit to your institution.

Contents

Part V Next Step: Deep Learning

Part I
Basic Concepts of Machine Learning for Earth Scientists

Chapter 1
Introduction to Machine Learning

1.1 Machine Learning: Definitions and Terminology

Shai and Shai (2014) define machine learning (ML) as "the automated detection of meaningful patterns in data." Since this is a broad definition, I am going to narrow it down by providing additional definitions from various authors (e.g., Samuel, 1959; Jordan & Mitchell, 2015; Géron, 2017; Murphy, 2012).

As example, Murphy (2012) defines ML as "the application of algorithms and methods to detect patterns in large data sets and the use of these patterns to predict future trends, to classify, or to make other types of strategic decisions."

In one of the earliest attempts to define ML, Samuel (1959) outlined one of the primary goals as "a computer that can learn how to solve a specific task, without being explicitly programmed." We can also take advantage of a more formal definition by Mitchell (1997): "A computer program is said to learn from experience E with respect to some task T and some performance measure P if its performance on T, as measured by P, improves with experience E." But what is "experience" for a computer program? In the physical sciences, experience for a computer program almost always coincides with data, so we can reword the definition by Mitchell (1997) to "A computer program is said to learn from data D with respect to some task T and some performance measure P if its performance on T, as measured by P, improves with the analysis of D."

One shared feature of ML methods is that they attempt to solve problems without requiring a detailed specification of the tasks to execute (Shai & Shai, 2014). Especiallyv when a human programmer cannot provide an explicit pathway to

Fig. 1.1 Artificial
intelligence, machine
learning, and deep learning

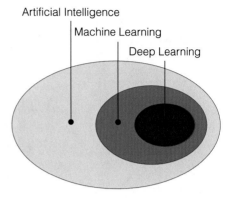

achieve the solution to the problem, these methods can often unravel the complexity of hidden patterns in the investigated data set and solve it.

By using set theory, we can define ML as a subset of artificial intelligence (AI), which is the effort to automate intellectual tasks normally performed by humans (Chollet, 2021) (Fig. 1.1). Note that AI covers a broad domain involving both ML and deep learning (DL). However, the AI set also includes numerous other approaches and techniques, some of which do not involve learning.

To summarize, the following are the key features of ML algorithms:

- ML methods try to extract meaningful patterns from a data set;
- ML algorithms are not explicitly programmed to solve a specific task;
- The learning process is a fundamental task in ML;
- ML methods learn from data;
- ML is a subset of AI;
- DL is a subset of ML.

When we start a new discipline, the first task is to learn the basic concepts and terminology. Table 1.1 gives a basic glossary to familiarize the geoscientist with the "language" used by data scientists, which is often difficult and sometimes misleading for a novice.

1.2   The Learning Process

As stated above, ML algorithms are not programmed to process a conceptual model defined a priori but instead attempt to uncover the complexities of large data sets through a so-called learning process (Bishop, 2007; Shai & Shai, 2014). In other words, the main goal of ML algorithms is to transform experience (i.e., data) into "knowledge" (Shai & Shai, 2014).

To better understand, we can compare the learning process of ML algorithms to that of humans. For example, humans begin learning to use the alphabet by observing the world around them where they find sounds, written letters, words, or phrases. Then, at school, they understand the significance of the alphabet and how to combine the different letters. Similarly, ML algorithms use the training data to learn significant patterns and then use the learned expertise to provide an output (Shai & Shai, 2014). One way to classify ML algorithms is by their degree of "supervision" (i.e., supervised, unsupervised, or semisupervised; Shai & Shai, 2014).

1.3 Supervised Learning

The training of supervised ML methods always provides both the input data and the desired solutions (i.e., the labels) to the algorithm. As an example, regression and classification tasks are suitable problems for supervised learning methods.

In classification tasks (Figs. 1.2a and b), ML algorithms try to assign a new observation to a specific class (i.e., a set of instances characterized by the same label) (Lee, 2019). If you do not understand some terms, please refer to Table 1.1. In regression problems (Fig. 1.2c and d), ML algorithms try, in response to an observation, to guess the value for one or more dependent variables.

Later in the book, we discuss extensively the application of regression and classification tasks in earth science problems (cf. Part III). However, Fig. 1.2 outlines two geological examples of supervised learning in the field of classification and

Table 1.1 Basic ML terminology. For a detailed glossary, please refer to the online ML course by Google™: https://bit.ly/mlglossary

Term	Description
Tensor	In ML, the word tensor typically describes a multidimensional array
Feature	An input variable used by ML algorithms
Attribute	Often used as a synonym feature
Label	Consists of the correct "answer" or "result" for a specific input tensor
Observation	A synonym for instance and example; a row of the data set, characterized by one or more features. In labeled data sets, observations also contain a label. In a geochemical data set, observations consist of one sample
Class	A set of observations characterized by the same label
Prediction	The output of a ML algorithm for a specific input observation
Model	What a ML algorithm has learned after training
Training a model	Process of determining the best model. Is is synonymous with the learning process
Training data set	The subset of the investigated data set used to train the model in the learning process
Validation data set	The subset of the investigated data set used to validate the model in the learning process
Test data set	An independent data set used to test the model after the validation process

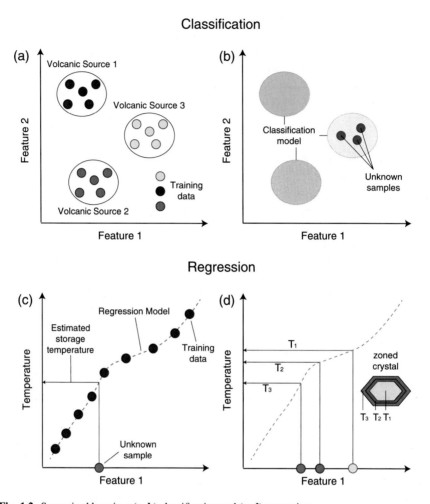

Fig. 1.2 Supervised learning: (**a**, **b**) classification and (**c**, **d**) regression

regression: (1) the identification of the volcanic source using glass shard compositions, which is a typical problem in tephrostratigraphy and tephrochronology (Lowe,
2011), and (2) the retrieval of magma storage temperatures based on clinopyroxene
chemistry (Petrelli et al., 2020).

1.4   Unsupervised Learning

Unsupervised learning acts on unlabeled training data. In other words, the ML
algorithm tries to identify significant patterns from the investigated data set without
the benefit of being fed external solutions. Fields that apply unsupervised learning

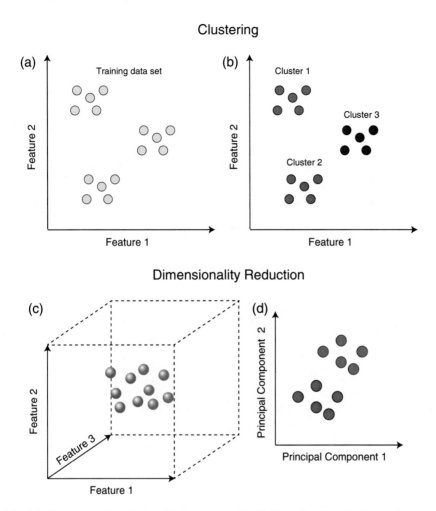

Fig. 1.3 Unsupervised learning: (**a, b**) clustering and (**c, d**) dimensionality reduction

include clustering, dimensionality reduction, and the detection of outliers or novelty observations.

Clustering consists of grouping "similar" observations into "homogeneous" groups (see Fig. 1.3a and b), which helps in discovering unknown patterns in unlabeled data sets. In the Earth Sciences, clustering has widespread applications in seismology (e.g., Trugman and Shearer, 2017), remote sensing (e.g., Wang et al., 2018), volcanology (e.g., Caricchi et al., 2020), and geochemistry (e.g., Boujibar et al., 2021) to cite a few.

The reduction of the dimensionality (Fig. 1.3c and d) of a problem reduces the number of features to treat, allowing the visualization of high-dimensional data sets (e.g., Morrison et al., 2017) or increasing the efficiency of a ML workflow.

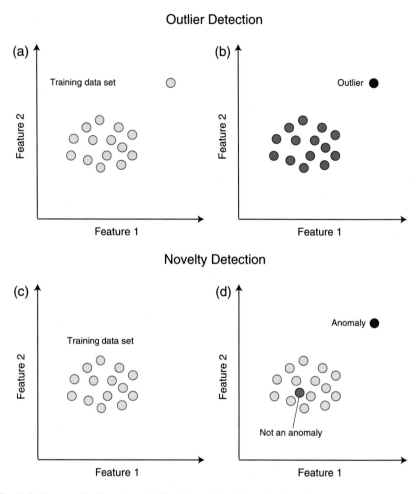

Fig. 1.4 Unsupervised learning: (**a, b**) outlier and (**c, d**) novelty detection

Tenenbaum et al. (2000) provide a concise but effective definition of dimensionality reduction: "finding meaningful low-dimensional structures hidden in their high-dimensional observations."

Finally, the detection of outlier or novelty observations (Fig. 1.4) deals with deciding whether a new observation belongs to a single set (i.e., an inlier) or should be considered different (i.e., an outlier or a novelty). The main difference between outlier and novelty detection lies in the learning process. In outlier detection (Fig. 1.4a and b), training data contain both inliers and potential outliers. Therefore, the algorithm tries to define which observation deviates from the others. In novelty detection (Fig. 1.4c and d), the training data set contains inliers only, and the algorithm tries to determine if a new observation is an outlier (i.e., a novelty).

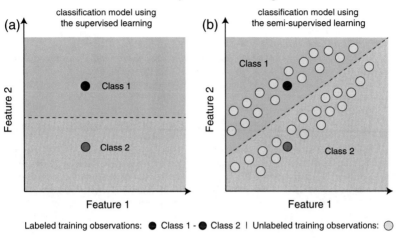

Fig. 1.5 (a) A supervised classification model using two labeled observations as the training data set. (b) A semisupervised classification model using the same two labeled observations from panel (a) plus many unlabeled instances

1.5 Semisupervised Learning

As you may argue, semisupervised learning falls somehow between supervised and unsupervised training methods. Typically, semisupervised algorithms learn from a small portion of labeled data and a large portion of unlabeled data (Zhu & Goldberg, 2009). More specifically, semisupervised learning algorithms use unlabeled data to improve supervised learning tasks when the labeled data are scarce or expensive (Zhu & Goldberg, 2009). To better understand, please see Fig. 1.5. In detail, Fig. 1.5a reports the results of a supervised classification model that uses two labeled observations as the training data set. Also, Fig. 1.5b displays a classification model resulting from semisupervised learning from the same two labeled data sets of Fig. 1.5a, plus several unlabeled observations.

References

Bishop, C. (2007). *Pattern recognition and machine learning.* Springer Verlag.
Boujibar, A., Howell, S., Zhang, S., Hystad, G., Prabhu, A., Liu, N., Stephan, T., Narkar, S., Eleish, A., Morrison, S. M., Hazen, R. M., & Nittler, L. R. (2021). Cluster analysis of presolar silicon carbide grains: Evaluation of their classification and astrophysical implications. *The Astrophysical Journal. Letters, 907*(2), L39. https://doi.org/10.3847/2041-8213/ABD102

Caricchi, L., Petrelli, M., Bali, E., Sheldrake, T., Pioli, L., & Simpson, G. (2020). A data driven approach to investigate the chemical variability of clinopyroxenes from the 2014–2015 Holuhraun–Bárdarbunga eruption (Iceland). *Frontiers in Earth Science, 8*. https://doi.org/10.3389/feart.2020.00018

Chollet, F. (2021). *Deep learning with Python.* (2nd ed.). Manning.

Geéron, A. (2017). *Hands-on machine learning with Scikit-Learn and TensorFlow: Concepts, tools, and techniques to build intelligent systems.* O'Reilly Media, Inc.

Jordan, M., & Mitchell, T. (2015). Machine learning: Trends, perspectives, and prospects. *Science, 349*(6245), 255–260. https://doi.org/10.1126/science.aaa8415

Lee, W.-M. (2019). *Phyton machine learning.* John Wiley & Sons Inc.

Lowe, D. J. (2011). Tephrochronology and its application: A review. *Quaternary Geochronology, 6*(2), 107–153. https://doi.org/10.1016/j.quageo.2010.08.003

Mitchell, T. M. (1997). *Machine Learning.* McGraw-Hill.

Morrison, S., Liu, C., Eleish, A., Prabhu, A., Li, C., Ralph, J., Downs, R., Golden, J., Fox, P., Hummer, D., Meyer, M., & Hazen, R. (2017). Network analysis of mineralogical systems. *American Mineralogist, 102*(8), 1588–1596. https://doi.org/10.2138/am-2017-6104CCBYNCND

Murphy, K. P. (2012). *Machine learning: A probabilistic perspective.* The MIT Press.

Petrelli, M., Caricchi, L., & Perugini, D. (2020). Machine learning thermo-barometry: Application to clinopyroxene-bearing magmas. *Journal of Geophysical Research: Solid Earth, 125*(9). https://doi.org/10.1029/2020JB020130

Samuel, A. L. (1959). Some studies in machine learning using the game of checkers. *IBM Journal of Research and Development, 3*, 210–229.

Shai, S.-S., & Shai, B.-D. (2014). *Understanding machine learning: From theory to algorithms.* Cambridge University Press.

Tenenbaum, J. B., De Silva, V., & Langford, J. C. (2000). A global geometric framework for nonlinear dimensionality reduction. *Science, 290*(5500), 2319–2323. https://doi.org/10.1126/SCIENCE.290.5500.2319

Trugman, D., & Shearer, P. (2017). GrowClust: A hierarchical clustering algorithm for relative earthquake relocation, with application to the Spanish Springs and Sheldon, Nevada, earthquake sequences. *Seismological Research Letters, 88*(2), 379–391. https://doi.org/10.1785/0220160188

Wang, Q., Zhang, F., & Li, X. (2018). Optimal clustering framework for hyperspectral band selection. *IEEE Transactions on Geoscience and Remote Sensing, 56*(10), 5910–5922. https://doi.org/10.1109/TGRS.2018.2828161

Zhu, X., & Goldberg, A. B. (2009). *Introduction to semi-supervised learning.* Morgan; Claypool Publishers.

Chapter 2
Setting Up Your Python Environments for Machine Learning

2.1 Python Modules for Machine Learning

The development of A ML model in Python uses both general-purpose scientific libraries (e.g., NumPy, ScyPy, and pandas) and specialized modules (e.g., scikit-learn,[1] PyTorch,[2] and TensorFlow[3]).

Scikit-Learn Scikit-learn is a Python module that solves small- to medium-scale ML problems (Pedregosa et al., 2011). It implements a wide range of state-of-the-art ML algorithms, making it one of the best options to start learning ML (Pedregosa et al., 2011).

PyTorch PyTorch is a Python package that combines high-level features for tensor management, neural network development, autograd computation, and back-propagation (Paszke et al., 2019). The PyTorch library grows within Meta's AI[4] (formerly Facebook AI) research team. In addition, it benefits from a strong ecosystem and a large user community that supports its development (Papa, 2021).

TensorFlow TensorFlow began at Google, and it was open-sourced in 2015. It combines tools, libraries, and community resources to develop and deploy DL models in Python (Bharath & Reza Bosagh, 2018).

---

[1] https://scikit-learn.org.

[2] https://pytorch.org.

[3] https://www.tensorflow.org.

[4] https://ai.facebook.com.

M. Petrelli, *Machine Learning for Earth Sciences*, Springer Textbooks
in Earth Sciences, Geography and Environment,
https://doi.org/10.1007/978-3-031-35114-3_2

2.2   A Local Python Environment for Machine Learning

The Individual Edition of the Anaconda Python Distribution[5] provides an example of a "ready-to-use" scientific Python environment to perform basic ML tasks with the scikit-learn module. It also allows advanced tasks such as installing libraries that are specifically developed for DL (e.g., PyTorch and TensorFlow). To install the Individual Edition of the Anaconda Python distribution, I suggest following the directives given in the official documentation.[6]

First, download and run the most recent stable installer for your operating system (i.e., Windows, Mac, or Linux). For Windows or Mac users, a graphical installer is also available. The installation procedure using the graphical installer is the same as for any other software application. The Anaconda installer automatically installs the Python core and Anaconda Navigator, plus about 250 packages defining a complete environment for scientific visualization, analysis, and modeling. Over 7500 additional packages, including PyTorch and TensorFlow, can be installed individually as needed from the Anaconda repository with the "conda"[7] package-management system. The basic tools to start learning and developing small- to medium-scale ML projects are the same as those used for any scientific Python Scientific project. Consequently, I suggest using Spyder and JupyterLab.

Spyder[8] is an integrated development environment that combines a text editor to write code, inspection tools for debugging, and interactive Python consoles for code execution (Fig. 2.1).

JupyterLab[9] is a web-based development environment to manage Jupyter Notebooks (i.e., web applications for creating and sharing computational documents, see Fig. 2.2)

2.3   ML Python Environments on Remote Linux Machines

Accessing and working on remote computational infrastructure is mandatory for large-scale and data-intensive ML workflows. However, the scope of the present book does not include providing a detailed description of how to develop high-performance computational infrastructure. Suffice it to say that such infrastructure often constitutes a cluster of Linux instances (i.e., virtual computing environments based on the Linux operating system), so we limit ourselves to describing how to connect to and work with a remote Linux instance. The present section shows how

---

[5] https://www.anaconda.com.

[6] https://www.anaconda.com/products/individual/.

[7] https://docs.conda.io/.

[8] https://www.spyder-ide.org.

[9] https://jupyter.org.

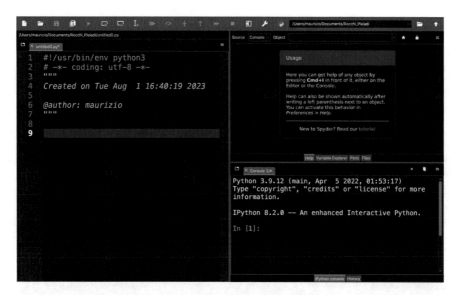

Fig. 2.1 Screenshot of Spyder integrated development environment. The text editor for writing code is on the left. The bottom-right panel is the IPython interactive console, and the top-right panel is the variable explorer

to set up a Debian instance on the Amazon Web Services™ (AWS) facilities. Next, it shows how to set up the Anaconda Individual Edition Python environment on your AWS Debian instance.

Figure 2.3 shows the Amazon management console of the "Elastic Compute Cloud" (EC2).[10] From the EC2 management console, a new computational instance can be launched by clicking the "Launch new instance" button. A guided step-by-step procedure follows. The user defines each detail of their computational instance [i.e., (1) chose the Amazon Machine Image; (2) choose the instance type, (3) define the key pair; further configure the instance, add storage, add tags, configure security group, and (4) launch the instance]. In steps (1–4) (see Fig. 2.4), I selected the Debian 10 64-bit (x86) Amazon Machine Image. Also, I selected the t2.micro instance type because it is eligible as a "free tier." Note that other options could be available as a "free tiers" and massive instance types could also be selected. As an example, the g5.48xlarge instance type consists of 192 virtual CPUs, 768 GiB of memory, and a network performance of 100 Gigabit. The total amount of computational power is only a matter of the budget at your disposal. The step 3 (see Fig. 2.5) consists of selecting an existing key pair or creating a new one. A "key pair" gives the security credentials to prove your identity when connecting to a remote instance. It consists of a "public key," which is stored in the remote instance, and a "private key," which is hosted in your machine. Anyone who possesses the private

---

[10] https://aws.amazon.com/ec2/.

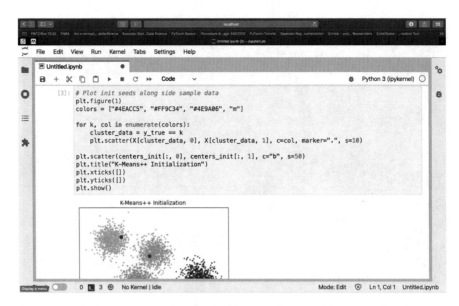

Fig. 2.2 Screenshot of Jupyter Notebook combining narrative text, code, and visualizations

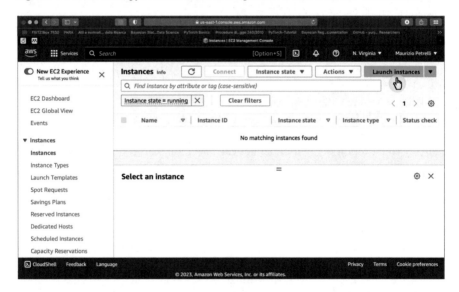

Fig. 2.3 Screenshot of the Elastic Compute Cloud (EC2) management console. The "Launch instance" button allows the user to start a new instance (April, 2023)

key of a specific key pair can connect to the instance that stores the associated public key. From your Linux and Unix OS (including the Mac OS), you can create a key pair by using the *ssh-keygen* command. However, the EC2 management console allows you to create and manage key pairs with a single click (Fig. 2.5). We can

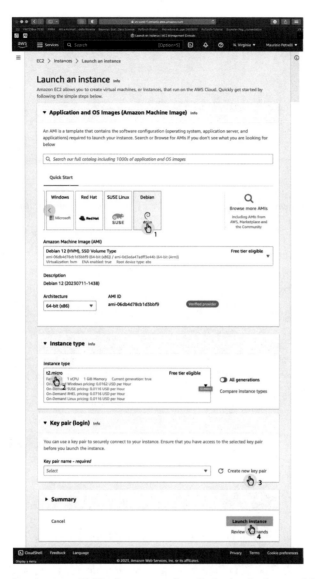

Fig. 2.4 Launch an instance: (1) The first step consists of selecting the Amazon Machine Image (AMI); (2) The second step consists of selecting the "Instance Type."; (3) Before launching a new instance you must select a "key pair;" (4) Finally, launch the instance (April, 2023)

safely set all the other instance parameters to their default values and click on the "Launch Instance" button.

The final step consists of launching the instance that, after initialization, appears in the EC2 management console (Fig. 2.6). To access an instance, select it in the EC2 management console and click on the "Connect" button (Fig. 2.6), which opens the

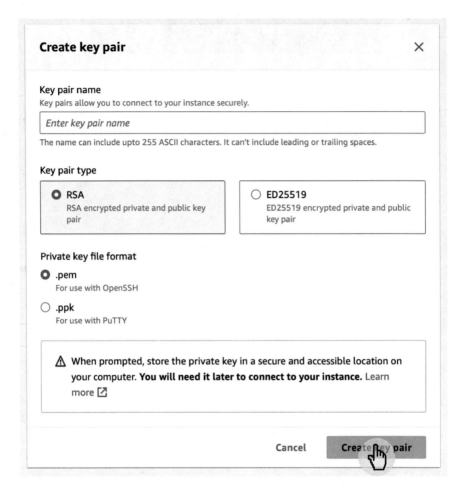

Fig. 2.5 How to create a "key pair" (April, 2023)

"Connect to instance" window, showing all available options to access the instance (Fig. 2.7). Our choice is to access the instance by using the Secure Shell (SSH) protocol (Fig. 2.7). The SSH Protocol is a cryptographic communication system for secure remote login and network services over an insecure network. It allows you to

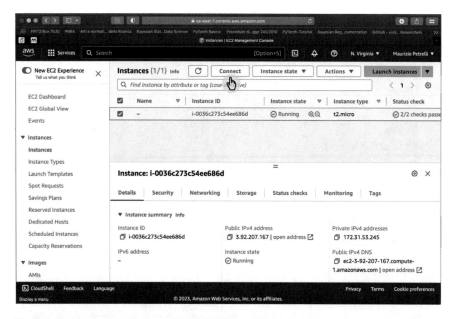

Fig. 2.6 Connecting to an instance (April, 2023)

"safely" connect and work on a remote instance from your desk or sofa. To connect to the remote instance, we need a SSH client (e.g., a Mac OS Terminal or PuTTY[11]) and into which we enter the following command:

```
ssh -i local_path/aws.pem user@user_name@host
```

where the *ssh* command initializes the SSH connection from the *user* account to the *host* (i.e., an IP or a domain name) remote instance. The -i option selects a specific private key (i.e., *aws.pem*) to pair with the public key in the *host* instance.

For the specific case shown in Fig. 2.7, I enter:

```
ssh -i /Users/maurizio/.ssh/aws.pem admin@ec2-52-91-26-146.
    compute-1.amazonaws.com
```

We are now connected to the remote instance in one AWS computing facility (Fig. 2.8) and we are ready to install the Anaconda Python Individual Edition from the command line.

Before starting the install procedure for the Anaconda Python Individual edition, I suggest upgrading the Debian packages as follows:

```
$ sudo apt-get update
$ sudo apt-get dist-upgrade
```

---

[11] https://www.putty.org.

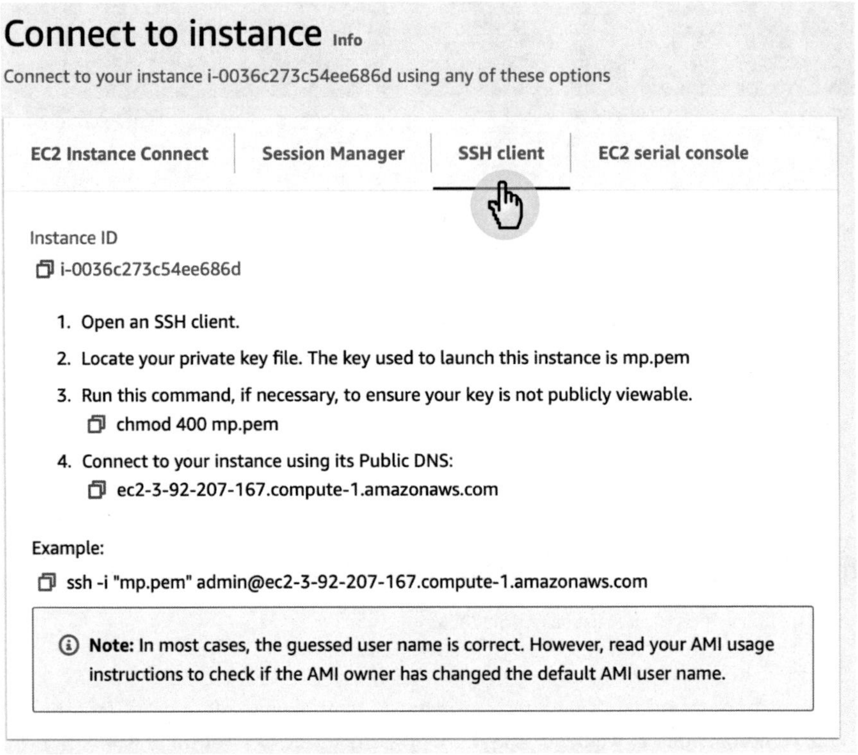

Connect to instance Info

Connect to your instance i-0036c273c54ee686d using any of these options

EC2 Instance Connect	Session Manager	SSH client	EC2 serial console

Instance ID

🗗 i-0036c273c54ee686d

1. Open an SSH client.
2. Locate your private key file. The key used to launch this instance is mp.pem
3. Run this command, if necessary, to ensure your key is not publicly viewable.
   🗗 chmod 400 mp.pem
4. Connect to your instance using its Public DNS:
   🗗 ec2-3-92-207-167.compute-1.amazonaws.com

Example:

🗗 ssh -i "mp.pem" admin@ec2-3-92-207-167.compute-1.amazonaws.com

> ⓘ **Note:** In most cases, the guessed user name is correct. However, read your AMI usage
> instructions to check if the AMI owner has changed the default AMI user name.

Fig. 2.7 Accessing by a SSH client (April, 2023)

The *sudo apt-get update* command gets you an updated list of packages. Then the *sudo apt-get* dist-upgrade will "intelligently" upgrade these packages, without upgrading the current Debian release. Now download the latest Anaconda Python distribution[12] for Linux-x86_64 using *curl*:

```
$ curl -O https://repo.anaconda.com/archive/Anaconda3-2023.03-
    Linux-x86_64.sh
```

if *curl* does not work, install it as follows:

```
$ sudo apt-get install curl
```

At this point, we need to verify the data integrity of the installer with cryptographic hash verification through the SHA-256 checksum. We use the *sha256sum* command along with the filename of the script:

```
$ sha256sum Anaconda3-2023.03-Linux-x86_64.sh
```

---

[12] https://repo.anaconda.com/archive/.

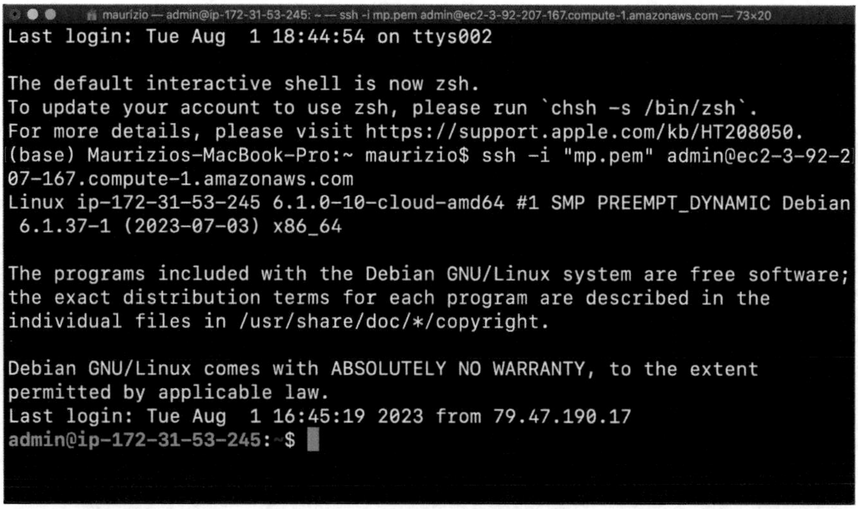

Fig. 2.8 Well done! You are connected to your remote instance

The result is

```
19737d5c27b23a1d8740c5cb2414bf6253184ce745d0a912bb235a212a15e075
```

and must match the cryptographic hash verification code in the Anaconda repository.[13] As a final step, we run the installation script:

```
$ bash Anaconda3-2023.03-Linux-x86_64.sh
```

It starts a step-by-step guided procedure starting with

```
Welcome to Anaconda3 py310_2023.03-0

In order to continue the installation process, please review the
    license
agreement.
Please, press ENTER to continue
```

Press "ENTER" to access the license information and continue clicking "ENTER" until you get the following question:

```
Do you approve the license terms? [yes|no]
```

---

[13] https://docs.anaconda.com/anaconda/install/hashes/lin-3-64/.

Type "yes" to get to the next step, which is the selection of the location for the installation:

```
Anaconda3 will now be installed into this location:
/home/admin/anaconda3

  - Press ENTER to confirm the location
  - Press CTRL-C to abort the installation
  - Or specify a different location below
```

I suggest pressing "ENTER" to retain the default location. At the end of the installation, you receive the following output:

```
...
installation finished.
Do you wish the installer to initialize Anaconda3
by running conda init? [yes|no]
[no] >>>
```

Type "yes". For changes to take effect, close and re-open the shell. Now, the base conda environment, highlighted by (base) at the beginning of the prompt command, should be active:

```
(base) [ec2-user@ip-172-31-35-226 ~]$
```

The base environment for ML in Python is now ready for use in your remote instance.

2.4 Working with Your Remote Instance

Once you are connected to your remote instance, for example, by

```
$ ssh -i local_path/aws.pem user@user_name@host
```

knowledge of the basic Linux OS commands is mandatory. However, a detailed explanation of the architecture, commands, and operations of the Linux OS is again beyond the scope of this book. Consequently, I suggest reading specialized books (Ward, 2021; Negus, 2015) to acquire the requisite skills. Table 2.1 lists common commands that allow you to transfer files between a local machine and remote instances. In addition, it provides basic tools for file management in a Linux environment.

To copy a file from your local machine to the remote instance and *vice versa* I suggest using the scp command, which is based on the SSH protocol. Specifically, the command is

```
$ scp  -i local_path/aws.pem filename user@host:/home/user/
    filename
```

Table 2.1 Basic Linux commands

Command	Description
ls	View the contents of a directory
cd..	Move one directory up
cd folder_name	Go to the folder named folder_name
cp myfile.jpg /new_folder	Copy myfile.jpg to the new_folder path
mv	Use mv to move files, the syntax is similar to cp
mkdir my_folder	Create a new folder named my_folder
rm	Delete directories and the contents within them (take care with rm!)
tar	Archive multiple files into a compressed file
chmod	Change the read, write, and execute permissions of files and directories
top	Display a list of running processes, CPU usage, and memory usage
pwd	Print the current working directory (i.e., the directory in which you are working)
sudo	Ii is the abbreviation of "SuperUser Do." It enables you to run tasks requiring administrative permissions. Take great care with sudo!

This command copies the file named "filename" from the local machine to the folder */home/user/* of the remote instance *host*. As explained in Sect. 2.3), the aws.pem private key stores the credentials to securely login to the *host instance*. To copy a file from your remote instance to the local machine use

```
$ scp  -i local_path/aws.pem  user@host:/home/user/filename /
    localfolder/filename
```

Finally, to launch a Python script we use the *python* command:

```
$ python myfile.py
```

To run multiple Python files you could use a bash script, which is a text file named *my_bash_script.sh*, and then run it as follows:

```
$ bash my_bash_script.sh
```

Here are two examples:

```
#!/bin/bash
/home/path_to_script/script1.py
/home/path_to_script/script2.py
/home/path_to_script/script3.py
/home/path_to_script/script4.py
```

and

```
#!/bin/bash
/home/path_to_script/script1.py &
/home/path_to_script/script2.py &
/home/path_to_script/script3.py &
/home/path_to_script/script4.py &
```

to run them sequentially and in parallel, respectively.

Note that the Anaconda Individual Edition comes with scikit-learn as a default package. DL packages such as Tensorflow and PyTorch must be installed separately. To avoid conflicts, I suggest creating isolated Python environments to work separately with PyTorch and TensorFlow.

2.5 Preparing Isolated Deep Learning Environments

Conda is an open-source package-management system and environment-management system developed by Anaconda[14] and that serves to install and update Python packages and dependencies. It also serves to manage isolated Python environments to avoid conflicts. As an example, consider the following statement:

```
conda create --name env_ml python=3.9 spyder scikit-learn
```

This statement creates a new Python 3.9 environment named env_ml with spyder, scikit-learn, and related dependencies installed. To activate the env_ml environment:

```
conda activate env_ml
```

to deactivate the current environment, use

```
conda deactivate
```

To list the available environments, use

```
conda info --envs
```

In the resulting list, the active environment is highlighted by *. Also, the active environment is usually given at the beginning of the command prompt [e.g., (base)]:

```
(base) admin@ip-172-31-59-186:~$
```

To remove an environment, use

```
conda remove --name env_ml --all
```

The following statement

```
conda env export > env_ml.yml
```

---

[14] https://www.anaconda.com/.

exports all information about the active environment to a file named env_ml.yml, which can then be used to share the environment to allow others to install it by using the following command:

```
conda env create -f env_ml.yml
```

More details on environment management are available in the conda official documentation.[15] The following listing resumes all the steps involved in creating a ML environment with DL functionalities based on PyTorch:

```
$ conda create --name env_pt python=3.9 spyder scikit-learn
$ conda activate env_pt
(env_pt)$ conda install pytorch torchvision torchaudio -c pytorch
```

The last command installs PyTorch, working on the CPU only, on my Mac. To find the right command for your hardware and operating system, please refer to the PyTorch website.[16]

Similarly, to create a ML environment based on scikit-learn with Tensorflow DL functionalities, use the following command:

```
$ conda create --name env_tf --channel=conda-forge tensorflow
```

As you can see, I used a specific channel (i.e., conda-forge[17]) to download tensorflow and spyder. Listing my conda environment now gives

```
$ conda info --envs
Output:
# conda environments:
#
base                     * /opt/anaconda3
env_ml                     /opt/anaconda3/envs/env_ml
env_pt                     /opt/anaconda3/envs/env_pt
env_tf                     /opt/anaconda3/envs/env_tf
```

2.6   Cloud-Based Machine Learning Environments

With cloud-based ML environments, I refer to Jupyter Notebook-based services, which are hosted in the cloud. Examples are Google™ Colaboratory, Kaggle, and Saturn Cloud. The first two services, Google™ Colaboratory and Kaggle, are both managed by Google™ and offer a free plan with limited computational resources. Finally, Saturn Cloud offers a free plan with 30 hours of computation. All services allow the online use of Jupyter Notebooks.

---

[15] https://docs.conda.io/.

[16] https://pytorch.org/get-started/locally/.

[17] https://conda-forge.org.

Fig. 2.9 Google™ Colaboratory (April, 2023)

Figures 2.9 and 2.10 provide a quick look at the entry-level notebooks for Google™ Colaboratory and Kaggle, respectively. Also, Figs. 2.9 and 2.10 show that both Google™ Colaboratory and Kaggle all come with scikit-learn, Tensorflow, and PyTorch installed and ready to use. Using Saturn Cloud™, a new Python Server can be launched by clicking the "New Python Server" server button (see Fig. 2.11), which opens a new window where you can personalize the instance. Note that the default configuration does not include either PyTorch or Tensorflow, although they can be added quickly in the "Extra Packages" section (Fig. 2.12). As an example, Fig. 2.12 shows how to add PyTorch. Finally, Fig. 2.13 demonstrates that the resulting environment comes with both scikit-learn and PyTorch.

Although all the reported cloud-based ML Jupyter environments are robust and flexible solutions, I suggest using Google™ Colaboratory or Saturn Cloud™ for novices.

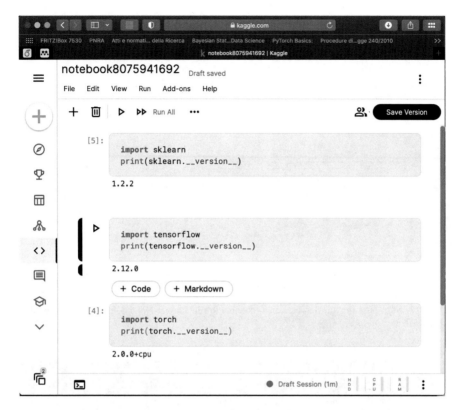

Fig. 2.10   Kaggle (April, 2023)

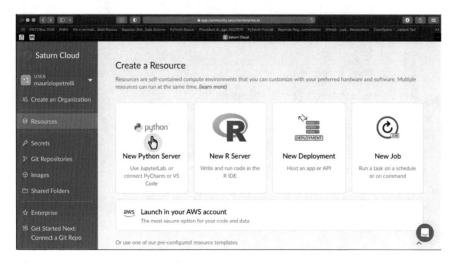

Fig. 2.11   Saturn Cloud™ (April, 2023)

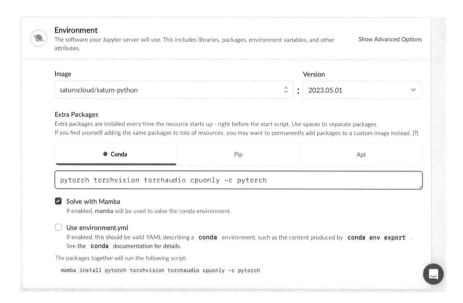

Fig. 2.12 Starting a Jupyter Server, i.e., a machine to run Jupyter Notebooks, in Saturn Cloud™ (April, 2023)

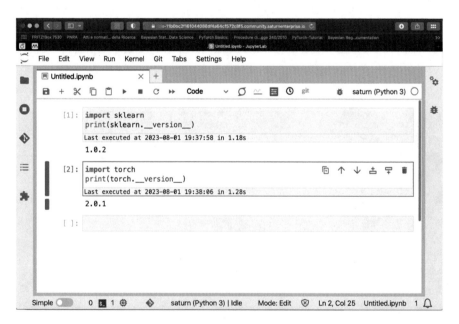

Fig. 2.13 Running a Jupyter Notebook in Saturn Cloud™ (April, 2023)

2.7   Speed Up Your ML Python Environment

A common argument by Python detractors is that Python is slow when compared with other established programming languages such as C or FORTRAN. We all agree with this argument but, in my opinion, this is not the point. In scientific computations, Python relies on libraries developed in higher-performing languages, mainly C and C++, and on parallel computing platforms such as CUDA.[18] For example, NumPy, the core Python library for scientific computing, is based on an optimized C code.[19] For ML purposes, all scikit-learn, PyTorch, and Tensorflow provide a base version of the library that can be safely installed in any local machine for rapid prototyping and small- to mid-scale problems. In addition, optimized versions for intensive computing applications are also available. For example, the Intel™ extension for scikit-learn accelerates ML applications in Python for Intel-based hardware by a factor 10–$100\times$.[20] The Intel™ extension for scikit-learn is easily installed by using *conda*. To prevent conflicts, I strongly recommend creating a new conda environment such as *env_ml_intel*:

```
$ conda create -n env_ml_intel -c conda-forge python=3.9 scikit-
    learn-intelex scikit-learn rasterio matplotlib pandas spyder
    scikit-image seaborn
```

Listing my local environments now gives

```
$ conda info --envs
Output:
# conda environments:
#
base                     * /opt/anaconda3
env_ml                     /opt/anaconda3/envs/env_ml
env_pt                     /opt/anaconda3/envs/env_pt
env_tf                     /opt/anaconda3/envs/env_tf
env_ml_intel               /opt/anaconda3/envs/env_ml_intel
```

I left the *base* environment untouched. Then I created two general-purpose ML environments, *env_ml* and *env_ml_intel*, with the latter optimized by Intel. Finally, I created two DL environments *env_pt* and *env_tf*, which are based on PyTorch and Tensorflow, respectively.

Note that DL libraries such as PyTorch and Tensorflow are highly optimized to support GPU computing (e.g., CUDA[21] and ROCm[22]). For example, a Pytorch CUDA-optimized version for the Linux OS can be easily installed by conda as follows (April, 2023):

---

[18] https://developer.nvidia.com/cuda-zone.

[19] https://numpy.org.

[20] https://github.com/intel/scikit-learn-intelex.

[21] https://developer.nvidia.com/cuda-zone.

[22] https://rocmdocs.amd.com/en/latest/.

```
$ conda install pytorch torchvision torchaudio pytorch-cuda=11.8
  -c pytorch -c nvidia
```

As already stated, providing a complete description of how to implement high-performance computing ML applications in Python is beyond the scope of this book. Therefore, please refer to the official documentation of each tool to get further details.

References

Bharath, R., & Reza Bosagh, Z. (2018). *TensorFlow for deep learning*. O'Reilly.

Negus, C. (2015). *Linux Bible* (9th ed., Vol. 112). John Wiley & Sons, Inc.

Papa, J. (2021). *PyTorch pocket reference*. O'Reilly Media, Inc.

Paszke, A., Gross, S., Massa, F., Lerer, A., Bradbury, J., Chanan, G., Killeen, T., Lin, Z., Gimelshein, N., Antiga, L., Desmaison, A., Köpf, A., Yang, E., DeVito, Z., Raison, M., Tejani, A., Chilamkurthy, S., Steiner, B., Fang, L., et al. (2019). PyTorch: An imperative style, high-performance deep learning library. In *Advances in neural information processing systems, 32*.

Pedregosa, F., Varoquaux, G. G., Gramfort, A., Michel, V., Thirion, B., Grisel, O., Blondel, M., Prettenhofer, P., Weiss, R., Dubourg, V., Vanderplas, J., Passos, A., Cournapeau, D., Brucher, M., Perrot, M., & Duchesnay, É. (2011). Scikit-learn: Machine learning in Python. *Journal of Machine Learning Research, 12*, 2825–2830.

Ward, B. (2021). *How Linux works, 3rd Edition: What every superuser should know*. No Starch Press, Inc.

Chapter 3
Machine Learning Workflow

3.1 Machine Learning Step-by-Step

Figure 3.1 shows a generalized workflow that is common to most ML projects. The first step is obtaining the data. In Earth Sciences, data can come from large-scale geological or geochemical samplings, remote-sensing platforms, well log analyses, or petrological experiments, to cite a few sources. The second step is pre-processing, which consists of all the operations required to prepare your data set for the successive steps of training and validation. Training the model involves running ML algorithms, which is the core business of a ML workflow. The validation step checks the quality of the training and ensures that the model is generalizable. Steps 3 and 4 are often closely connected and iterated many times to improve the quality of the results. Finally, the last step consists of deploying and securing your model.

We shall now evaluate each step and provide insights into how to successfully run a ML model in the field of Earth Sciences.

3.2 Get Your Data

Your data set repository may have many different formats. The easiest data sets consist of tabular data stored in text (e.g., .csv) or Excel™ files. Sometimes, a Structured Query Language (SQL) database hosts your data. Larger data sets may be stored in the Hierarchical Data Format (HDF5),[1] Optimized Row Columnar

© The Author(s), under exclusive license to Springer Nature Switzerland AG 2023

M. Petrelli, *Machine Learning for Earth Sciences*, Springer Textbooks
in Earth Sciences, Geography and Environment,
https://doi.org/10.1007/978-3-031-35114-3_3

Fig. 3.1 Workflow of a ML model

Table 3.1 Pandas methods to import standard and *state-of-the-art* file formats for ML applications

Method	Description	Comment
read_table()	Read general delimited file	Slow, not for large data sets
read_csv()	Read comma-separated values (csv) files	Slow, not for large data sets
read_excel()	Read Excel files	Slow, not for large data sets
read_sql()	Read sql files	Slow, not for large data sets
read_pickle()	Read pickled objects	Fast, not for large data sets
read_hdf()	Read Hierarchical Data Format (HDF) files	Fast, good for large data sets
read_feather()	Read feather files	Fast, good for large data sets
read_parquet()	Read parquet files	Fast, good for large data sets
read_orc()	Read Optimized Row Columnar files	Fast, good for large data sets

(ORC),[2] Feather (i.e., Arrow IPC columnar format),[3] or Parquet Format,[4] to cite a few.

For data that fit into your random access memory (RAM), pandas is probably the best choice for data import and manipulation (e.g., slicing, filtering) through *DataFrames*. Table 3.1 describes the potential of pandas methods for input and output (I\O).

If the data set starts filling your RAM entirely, Dask[5] is the probably the library of choice to manage your data and scale your Python code to parallel environments. Dask is a library designed to deal with "Big Data" through parallel computing in Python. Dask extends the concept of *DataFrames* to Dask DataFrames, which are large parallel *DataFrames* composed of many smaller pandas *DataFrames*. We introduce Dask and parallel computing later in Part IV of the book Before that, however, we must import our data sets for Earth Sciences ML applications using

---

[1] https://www.hdfgroup.org/solutions/hdf5/.

[2] https://orc.apache.org.

[3] https://arrow.apache.org/docs/python/feather.html.

[4] https://parquet.apache.org.

[5] https://dask.org.

pandas (see code listing 3.1). The investigated data set is available for download from the website[6] of the Laboratory for Space Sciences, Physics Department, Washington University in St. Louis. It deals with presolar SiC grains, extracted from meteorites (Stephan et al., 2021).

```
1  import pandas as pd
2
3  my_data = pd.read_excel("PGD_SiC_2021-01-10.xlsx", sheet_name='
       PGD-SIC')
4  print(my_data.info(memory_usage="deep"))
5
6  '''
7  Output:
8  <class 'pandas.core.frame.DataFrame'>
9  RangeIndex: 19978 entries, 0 to 19977
10 Columns: 123 entries, PGD ID to err[d(138Ba/136Ba)]
11 dtypes: float64(112), object(11)
12 memory usage: 29.4 MB
13 '''
```

Listing 3.1 Importing an Excel data set in Python

I assume that you are familiar with the *read_excel* statement in pandas. If not, I strongly suggest that you start with an introductory book such as "Introduction to Python in Earth Science Data Analysis" (Petrelli, 2021). The statement at line 4 of code listing 3.1 tells you how much memory is required to store our data set. In this case, the imported data set, consisting of approximately 20 000 rows and 123 columns, requires 24.4 MB, which is far less than the 32 GB of my MacBook™ Pro.

Large data sets [i.e., approaching or exceeding tera (10^{12}) or peta (10^{15}) bytes] cannot be efficiently stored in text files such as .csv files or in Excel files. Standard relational databases such as PostgreSQL, MySQL, and MS-SQL can store large quantities of information but are inefficient (i.e., too slow) compared with *state-of-the-art* high-performance data software libraries and file formats for managing, processing, and storing huge amounts of data. The formal definition of Big Data proposed by De Mauro et al. (2016) covers the three concepts of volume, velocity, and variety: "Big Data is the Information asset characterized by such a High Volume, Velocity and Variety to require specific Technology and Analytical Methods for its transformation into Value." A detailed description of data storage and analysis frameworks for Big Data is beyond the scope of this book, so I suggest that those interested consult specialized texts (Pietsch, 2021; Panda et al., 2022). Herein, we simply compare the performances of pandas for writing and reading GB-scale .csv and .hdf files on a MacBook pro (2.3 GHz Quad-Core Intel Core i7, 32 GB RAM). For example, code listing 3.2 generates a pandas *DataFrame* of ≈ 10 GB named

---

[6] https://presolar.physics.wustl.edu/presolar-grain-database/.

my_data and composed of random numbers hosted in 26 columns and 5×10^7 rows.
I used *my_data.info(memory_usage = "deep")*, code listing 3.3, to check the real
memory use of *my_data*, which is 9.7 GB.

Code listing 3.4 shows the execution time required to write (In [1], In [2], and
In [3]) and read (In [4], In [5], and In [6]) from text (.csv), parquet, and hdf5 files,
respectively. The results show that saving a .csv file takes about 25 minutes, which
is quite a long time! In contrast, saving the parquet and hdf5 files take 7 and 12 s,
respectively. Reading times are of the same order of magnitude: about 5 minutes for
.csv and 30 s for parquet and hdf5 files.

```
1 import pandas as pd
2 import numpy as np
3 import string
4
5 my_data = pd.DataFrame(np.random.normal(size=(50000000, 26)),
6                         columns=list(string.ascii_lowercase))
```

Listing 3.2   Generating a mid-size data set of about 10 GB

```
1 In [1]: my_data.info(memory_usage="deep")
2 <class 'pandas.core.frame.DataFrame'>
3 RangeIndex: 50000000 entries, 0 to 49999999
4 Data columns (total 26 columns):
5  #    Column  Dtype
6 ---   ------  -----
7  0    a       float64
8  1    b       float64
9  2    c       float64
10 3    d       float64
11 4    e       float64
12 5    f       float64
13 6    g       float64
14 7    h       float64
15 8    i       float64
16 9    j       float64
17 10   k       float64
18 11   l       float64
19 12   m       float64
20 13   n       float64
21 14   o       float64
22 15   p       float64
23 16   q       float64
24 17   r       float64
25 18   s       float64
26 19   t       float64
27 20   u       float64
28 21   v       float64
```

```
29  22  w        float64
30  23  x        float64
31  24  y        float64
32  25  z        float64
33 dtypes: float64(26)
34 memory usage: 9.7 GB
```

Listing 3.3 Checking the memory usage of our *DataFrame*

In light of the evidence given by code listing 3.4, I would suggest discontinuing the use of text files to store and retrieve your data sets at GB or larger scales in favor of binary files such as hdf5 or parquet. The case for this becomes particularly strong once the data dimensions grow significantly.

```
 1 In [1]: %time my_data.to_csv('out.csv')
 2 CPU times: user 22min 48s, sys: 55.8 s, total: 23min 44s
 3 Wall time: 24min 16s
 4
 5 In [2]: %time my_data.to_parquet('out.parquet')
 6 CPU times: user 13.1 s, sys: 2.71 s, total: 15.8 s
 7 Wall time: 11.8 s
 8
 9 In [3]: %time my_data.to_hdf('out.h5', key="my_data", mode="w")
10 %time my_data.to_hdf('out.h5', key="my_data1", mode="w")
11 CPU times: user 39.2 ms, sys: 4.33 s, total: 4.37 s
12 Wall time: 6.59 s
13
14 In [4]: %time my_data_1 = pd.read_csv('out.csv')
15 CPU times: user 3min 28s, sys: 37.7 s, total: 4min 5s
16 Wall time: 4min 45s
17
18 In [5]: %time my_data1 = pd.read_parquet('out.parquet')
19 CPU times: user 12.7 s, sys: 26.3 s, total: 39 s
20 Wall time: 31 s
21
22 In [6]: %time my_data1 = pd.read_hdf('out.h5', key='my_data')
23 CPU times: user 10.2 s, sys: 12.7 s, total: 23 s
24 Wall time: 28.8 s
```

Listing 3.4 Performances of the pandas library in writing and loading .cvs, .parquet and .h5 files

3.3 Data Pre-processing

Pre-processing consists of all operations required to prepare your data set for the next steps (e.g., training and validation; Maharana et al., 2022). This step is crucial because it converts raw data into a form suitable to build a ML model. While developing a ML project, you will likely spend most of your time preparing your data for the training. In detail, pre-processing refers to preparing (e.g., cleaning, organizing, normalizing) the raw data before moving to the training. In addition, pre-processing includes the preliminary steps to allow validation (e.g., train-test splitting).

3.3.1 Data Inspection

Data inspection is the qualitative investigation of a data set and allows one to become familiarized with the data set. A fundamental task of data inspection is descriptive statistics, which provides a clear understanding of the "shape" and structure of the data. To see how descriptive statistics can help, consider the following example: By looking at the histogram distributions, you can start arguing whether methods that require specific assumptions (e.g., a Gaussian structure) are well suited to analyze your data.

Code listing 3.5 shows how to undertake a preliminary determination of the main descriptive indexes of location, such as the mean and the median (e.g., p_{50} or the 50% percentile), and dispersion, such as the standard deviation and range (e.g., $range = max - min$) or the interquartile range (e.g., $iqr = p_{75} - p_{25}$).

```
In [1]: sub_data = my_data[['12C/13C', '14N/15N']]

In [2]: sub_data.describe().applymap("{0:.0f}".format)

Out[2]:
         12C/13C 14N/15N
count    19581    2544
mean        66    1496
std        207    1901
min          1       4
25%         44     336
50%         55     833
75%         69    2006
max      21400   19023
```

Listing 3.5 Determining descriptive statistics in Python

Figure 3.2 and code listing 3.6 show how Python can be used to statistically visualize a data set. In more detail, Fig. 3.2 shows the distribution of data in the $^{14}N/^{15}N$ versus $^{12}C/^{13}C$ projection (left panel) and the histogram distribution of $^{12}C/^{13}C$ (right panel).

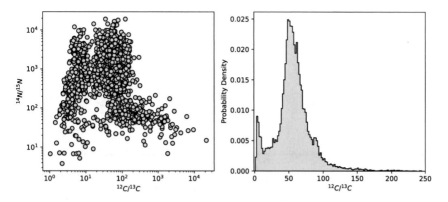

Fig. 3.2 Descriptive statistics (code listing 3.6)

```
1  import matplotlib.pyplot as plt
2
3  fig = plt.figure(figsize=(9,4))
4  ax1 = fig.add_subplot(1,2,1)
5  ax1.plot(my_data['12C/13C'], my_data['14N/15N'],
6              marker='o', markeredgecolor='k',
7              markerfacecolor='#BFD7EA', linestyle='',
8              color='#7d7d7d',
9              markersize=6)
10 ax1.set_yscale('log')
11 ax1.set_xscale('log')
12 ax1.set_xlabel(r'$^{12}C/^{13}C$')
13 ax1.set_ylabel(r'$^{14}N/^{15}N$')
14
15 ax2 = fig.add_subplot(1,2,2)
16 ax2.hist(my_data['12C/13C'], density=True, bins='auto',
17        histtype='stepfilled', color='#BFD7EA', edgecolor='
   black',)
18 ax2.set_xlim(-1,250)
19 ax2.set_xlabel(r'$^{12}C/^{13}C$')
20 ax2.set_ylabel('Probability Density')
21
22 fig.set_tight_layout(True)
```

Listing 3.6 Obtaining descriptive statistics in Python

3.3.2   Data Cleaning and Imputation

In real-world data sets such as geological data sets, "unwanted" entries are ubiquitous (Zhang, 2016). Examples include voids (i.e., missing data), "Not a Number" (NaN) entries, and large outliers. Cleaning a data set mainly consists of removing such unwanted entries. For example, the methods *.dropna()* and *.fillna()*

help when working with missing data; these are imported by pandas as NaN (see code listing 3.7).

```
 1  import pandas as pd
 2
 3  cleaned_data = my_data.dropna(
 4      subset=['d(135Ba/136Ba)', 'd(138Ba/136Ba)'])
 5
 6  print("Before cleaning: {} cols".format(my_data.shape[0]))
 7  print("After cleaning: {} cols".format(cleaned_data.shape[0]))
 8
 9  '''
10  Output:
11  Before cleaning: 19978 cols
12  After cleaning: 206 cols
13  '''
```

Listing 3.7 Removing NaN values

In detail, the *.dropna()* at line 3 removes all the rows where the isotopic value of $\delta^{135}Ba_{136}$ [‰] or $\delta^{138}Ba_{136}$ [‰] are missing.

Although appealing for its simplicity, removing entries containing missing values has some drawbacks, the most significant of which is the loss of information (Zhang, 2016). In particular, when dealing with a large number of features, a substantial number of observations may be removed because a single feature is missing, potentially introducing large biases (Zhang, 2016). A possible solution is data imputation, which is the replacement of missing values with imputed values. Several methods have been developed for data imputation, the easiest of which consists of replacing missing values with the mean, median, or mode of the investigated feature (Zhang, 2016). In pandas, *.fillna()* replaces NaN entries with text or a specific value. Also, the *SimpleImputer()* in scikit-learn imputes missing values with the mean, median, or mode.

A more evolved strategy consists of data imputation with regression (Zhang, 2016). In this case, you first fit a regression model (e.g., linear or polynomial) and then use the model to impute missing values (Zhang, 2016). In scikit-learn, the function *IterativeImputer()* develops an imputation strategy based on multiple regressions.

3.3.3   Encoding Categorical Features

Most available machine learning algorithms do not support the use of categorical (i.e., nominal) features. Therefore, categorical data must be encoded (i.e., converted to a sequence of numbers). In scikit-learn, *OrdinalEncoder()* encodes categorical features such as integers (i.e., 0 to $n_{categories} - 1$).

3.3.4   Data Augmentation

Data augmentation aims to increase the generalizability of ML models by increasing the amount of information in our data sets (Maharana et al., 2022), which consists of either adding modified copies of the available data (e.g., flipped or rotated images in the case of image classification) or combining existing features to generate new features. For example, Maharana et al. (2022) describe six data augmentation techniques for image analysis: (1) symbolic augmentation, (2) rule-based augmentation, (3) graph-structured augmentation, (4) mixup augmentation, (5) feature-space augmentation, and (6) neural augmentation (Maharana et al., 2022). Although the details of feature augmentation are far beyond the scope of this book, we will exploit data augmentation in Chap. 8 by following the strategy proposed by Bestagini et al. (2017).

3.3.5   Data Scaling and Transformation

The scaling and transformation of a data set is often a crucial step in ML workflows. Many ML algorithms strongly benefit from a preliminary "standardization" of the investigated data set. For example, all algorithms that use the Euclidean distance (and there are many of them!) as fundamental metrics may be significantly biased upon introducing features that differ significantly in magnitude.

Definition In a standardized data set, all features are centered on zero and their variance is of the same order of magnitude.

If a feature variance is orders of magnitude greater than the other feature variances, it might play a dominant role and prevent the algorithm from correctly learning the other features. The easiest way to standardize a data set is to subtract the mean and scale to unit variance:

$$\tilde{x}_e^i = \frac{x_e^i - \mu^e}{\sigma_s^e}. \tag{3.1}$$

In Eq. (3.1), $\tilde{x}_e^i$ and x_e^i are the transformed and original components, respectively. For example, they could belong to the sample distribution of a chemical element e such as SiO_2 or TiO_2 characterized by a mean μ^e and a standard deviation σ_s^e.

Scikit-learn implements Eq. (3.1) in the *sklearn.preprocessing.StandardScaler()* method.

In addition, scikit-learn implements additional scalers and transformers, which perform linear and nonlinear transformations, respectively. For example, *MinMaxScaler()* scales each feature belonging to a data set to a given range (e.g., between 0 and 1).

QuantileTransformer() provides nonlinear transformations that shrinks distances between marginal outliers and inliers, and *PowerTransformer*() provides nonlinear transformations in which data are mapped to a normal distribution to stabilize variance and minimize skewness.

The presence of outliers may affect the outputs of the model. If the data set has outliers, robust scalers or transformers are more appropriate. By default, *RobustScaler*() removes the median and scales the data according to the interquartile range. Note that *RobustScaler*() does not remove any of the outliers. Table 3.2 summarizes the main scalers and the transformers available in scikit-learn.

When the estimation uncertainties are quantified (e.g., by one sigma or one standard error), the data set could be cleaned to remove all data where the error exceeds a threshold of your choosing.

```python
import matplotlib.pyplot as plt
from sklearn.preprocessing import MinMaxScaler
from sklearn.preprocessing import StandardScaler
from sklearn.preprocessing import RobustScaler

X = my_data[['d(30Si/28Si)','d(29Si/28Si)']].to_numpy()

scalers = [("Unscaled", X),
           ("Standard Scaler", StandardScaler().fit_transform(X)),
           ("Min. Max. Scaler", MinMaxScaler().fit_transform(X)),
           ("Robust Scaler", RobustScaler().fit_transform(X))]

fig = plt.figure(figsize=(10,7))

for ix, my_scaler in enumerate(scalers):
    ax = fig.add_subplot(2,2,ix+1)
    scaled_X = my_scaler[1]
    ax.set_title(my_scaler[0])
    ax.scatter(scaled_X[:,0], scaled_X[:,1],
               marker='o', edgecolor='k', color='#db0f00',
               alpha=0.6, s=40)
    ax.set_xlabel(r'${\delta}^{30}Si_{28} [\perthousand]$')
    ax.set_ylabel(r'${\delta}^{29}Si_{28} [\perthousand]$')

fig.set_tight_layout(True)
```

Listing 3.8 Scalers and transformers

Finally, taking the logarithm of the data sometimes helps to reduce the skewness of the sample, assuming the data set follows a log-normal distribution (Limpert et al., 2001; Corlett et al., 1957). Code listing 3.8 shows how to apply various scalers and transformers to the log-transformed $^{12}C/^{13}C$ SiC data, and Fig. 3.3 shows the results.

Table 3.2 Scalers and transformers in scikit-learn. Descriptions are taken from the official documentation of scikit-learn

	Description
Scaler	
sklearn.preprocessing.StandardScaler()	Standardize features by removing the mean and scaling to unit variance [Eq. (3.1)]
sklearn.preprocessing.MinMaxScaler()	Transform features by scaling each feature to a given range. The default range is [0,1]
sklearn.preprocessing.RobustScaler()	Scale features using statistics that are robust against outliers. This scaler removes the median and scales the data according to the quantile range. The default quantile range is the interquartile range
Transformer	
sklearn.preprocessing.PowerTransformer()	Apply a power transform feature-wise to make data more Gaussian-like
sklearn.preprocessing.QuantileTransformer()	Transform features using quantile information. This method transforms features to follow a uniform or normal distribution. Therefore, for a given feature, this transformation tends to spread the most frequent values

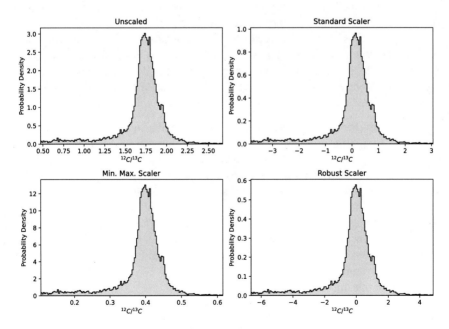

Fig. 3.3 Data sets scaled and transformed by code listing 3.8

3.3.6 *Compositional Data Analysis (CoDA)*

Before applying any statistical method, including ML algorithms, the underlying assumptions must be verified. An example is the assumption of normality, which is behind many methods. Other assumptions may regard the topology of the sample space. Geochemical determinations are an example of so-called compositional data (Aitchison, 1982; Aitchison & Egozcue, 2005; Razum et al., 2023), which are samples of non-negative multivariate data that are expressed relative to a fixed total (typically unity or percentages summing to 100%). The analysis of compositional data is called "compositional data analysis" (CoDA; Aitchison, 1984).

In compositional data, the sample space is represented by the Aitchison simplex s^D:

$$s^D = \left\{ \mathbf{x} = [x_1, x_2, x_i, \ldots, x_D] \mid x_i > 0, \ i = 1, 2, \ldots, D; \ \sum_{i=1}^{D} x_i = C \right\},$$

$$(3.2)$$

where C is a constant, typically 1 or 100. Compositional data typically share two characteristics: (1) the data are always positive and (2) the data sum to a constant (i.e., they are not independent). These characteristics hinder the application of many statistical methods because they often assume independent input samples in the interval $[-\infty, \infty]$. From the topological point of view, the simplex (i.e., the sample space for compositional vectors) differs radically from the Euclidean space associated with unconstrained data (Aitchison, 1982; Aitchison & Egozcue, 2005; Razum et al., 2023). Therefore, any method relying on the Euclidean distance should not be used directly with compositional data. Four established transformations are available that attempt to map the Aitchison simplex to Euclidean space.

Pairwise Log Ratio Transformation (*pwlr*) (Aitchison, 1982; Aitchison & Egozcue, 2005; Razum et al., 2023) The *pwlr* transformation maps a composition isometrically from a D-dimensional Aitchison simplex to a $D(D-1)/2$-dimensional space. In detail, it computes each possible log ratio but accounts for the fact that $\log(A/B) = -\log(B/A)$, so only one of them is needed. On data transformed by the pairwise log ratio, we can apply multivariate methods that do not rely on the invertibility of the covariance function. The interpretation of *pwlr*-transformed data is quite simple because each component results from a simple operation of division and is then transformed by a logarithm to reduce the skew of the resulting features.

The *pwlr* transformation is given by

$$pwlr(\mathbf{x}) = \left[\xi_{ij} \mid i < j = 1, 2, \ldots, D \right],$$

$$(3.3)$$

where $\xi_{ij} = \ln(x_i/x_j)$. Note that the redundancy of *pwlr* generates $D(D-1)/2$ features, which corresponds to an extremely-high-dimensional space.

Additive Log Ratio Transformation (*alr*) (Aitchison, 1982; Aitchison & Egozcue, 2005; Razum et al., 2023) The *alr* transformation is given by

$$alr(\mathbf{x}) = \left[\ln \frac{x_1}{x_D}, \ln \frac{x_2}{x_D}, \ldots, \ln \frac{x_{D-1}}{x_D} \right]. \tag{3.4}$$

The *alr* transformation nonisometrically maps vectors from the D-dimensional Aitchison simplex to a $(D-1)$-dimensional space.

As in the case of *pwlr*, the interpretation of *alr* data is quite simple because they also derive from a simple operation of division followed by a logarithm to reduce the skew of the resulting features.

Centered Log Ratio Transformation (*clr*) This transformation is given by

$$clr(\mathbf{x}) = \left[\ln \frac{x_1}{g(\mathbf{x})}, \ln \frac{x_2}{g(\mathbf{x})}, \ldots, \ln \frac{x_D}{g(\mathbf{x})} \right], \tag{3.5}$$

where $g_m(\mathbf{x})$ is the geometric mean of the parts of $\mathbf{x}$. The *clr* transformation isometrically maps the vectors from the D-dimensional Aitchison simplex to a D-dimensional Euclidean space. The *clr*-transformed data can then be analyzed by all multivariate tools that do not rely on a full rank of the covariance (Aitchison, 1982; Aitchison & Egozcue, 2005; Razum et al., 2023).

Orthonormal Log Ratio Transformation (*olr*) This transformation is also known as the isometric log ratio transformation (*ilr*). The *olr* coordinates of $\mathbf{x}$ with respect to the basis elements $\mathbf{e}_l$, $l = 1, 2, \ldots, n-1$, are defined as (Egozcue & Pawlowsky-Glahn, 2005)

$$x_l^* = \sqrt{\frac{rs}{r+s}} \ln \left[\frac{g(x_{k+1}, \ldots, x_{k+r})}{g(x_{k+r+1}, \ldots, x_{k+r+s})} \right], \tag{3.6}$$

where x_l^* is the *balance* between the groups of parts $x_{k+1}, \ldots, x_{k+r}$ and $g(x_{k+r+1}, \ldots, x_{k+r+s})$ and $\mathbf{e}_l$ is the *balancing element* for the two sets of parts (Egozcue & Pawlowsky-Glahn, 2005).

Note that "with defined balances, which are directly associated with an orthogonal coordinate system in the simplex, every multivariate statistical technique is available without any restriction and data can be properly statistically evaluated" (Razum et al., 2023). Each of the above-mentioned transformations is endowed with unique properties that can be used for compositional data analysis. The *clr* transformation is often used to construct compositional biplots and for cluster analysis (van den Boogaart & Tolosana-Delgado, 2013). Although *alr*-transformed data can be analyzed by using multivariate statistical tools, the *alr* transformation defines "coordinates in an oblique basis, something that affects distances if the usual Euclidean distance is computed from the *alr* coordinate" (van den Boogaart & Tolosana-Delgado, 2013). Consequently, the *alr* transformation "should not be used in cases [in which] distances, angles, and shapes are involved, as it deforms them" (Pawlowsky-Glahn & Buccianti, 2011). Any multivariate technique can be applied safely to *olr*-transformed data because it is related to the orthonormal basis of the simplex (Razum et al., 2023).

In Python, both scikit-bio[7] and pytolite[8] provide us with methods in the framework of CoDA.

3.3.7 A Working Example of Data Pre-processing

The code listings 3.9 and 3.10 show a step-by-step reproduction of data pre-processing by Boujibar et al. (2021) for a study of the clustering of pre-solar silicon carbide (SiC) grains. Do not worry if you cannot follow the specific cosmo-chemical problem investigated by Boujibar et al. (2021). The aim of the example is to highlight how to prepare a data set for ML investigations.

```
 1 import pandas as pd
 2 import matplotlib.pyplot as plt
 3 import numpy as np
 4 from sklearn.preprocessing import StandardScaler
 5 from sklearn.preprocessing import RobustScaler
 6
 7 # Import Data
 8 my_data = pd.read_excel("PGD_SiC_2021-01-10.xlsx",
 9                         sheet_name='PGD-SIC')
10
11 # limit to features of interest
12 my_data = my_data[['PGD ID', 'PGD Type', 'Meteorite', '12C/13C',
13                    'err+[12C/13C]', 'err-[12C/13C]', '14N/15N',
14                    'err+[14N/15N]', 'err-[14N/15N]',
15                    'd(29Si/28Si)', 'err[d(29Si/28Si)]',
16                    'd(30Si/28Si)', 'err[d(30Si/28Si)]']]
17
18 # Drop NaN
19 my_data = my_data.dropna()
20
21 # Removing M grains with large Si errors
22 my_data = my_data[~((my_data['err[d(30Si/28Si)]']>10) &
23                     (my_data['err[d(29Si/28Si)]']>10) &
24                     (my_data['PGD Type']== 'M'))]
25
26 # Excluding C and U grains
27 my_data = my_data[(my_data['PGD Type']=='X') |
28                   (my_data['PGD Type']=='N') |
```

---

[7] http://scikit-bio.org.

[8] https://pyrolite.readthedocs.io.

```
29                    (my_data['PGD Type']=='AB') |
30                    (my_data['PGD Type']=='M')  |
31                    (my_data['PGD Type']=='Y')  |
32                    (my_data['PGD Type']=='Z')]
33
34  # Excluding contaminated grains
35  my_data = my_data[~(((my_data['12C/13C']<93.56) &
36                      (my_data['12C/13C']>88.87)) &
37                     ((my_data['14N/15N']<339.94) &
38                      (my_data['14N/15N']>248)) &
39                     ((my_data['d(30Si/28Si)']<50)&
40                      (my_data['d(30Si/28Si)']>-50)) &
41                     ((my_data['d(29Si/28Si)']<50)&
42                      (my_data['d(29Si/28Si)']>-50))
43                    )]
```

Listing 3.9 Working example of data pre-processing (part 1)

```
1   # Trasform silica isotopic delta to isotopic ratios
2   Si29_28_0 = 0.0506331
3   Si30_28_0 = 0.0334744
4   my_data['30Si/28Si'] = ((my_data['d(30Si/28Si)']/1000)+1) *
        Si30_28_0
5   my_data['29Si/28Si'] = ((my_data['d(29Si/28Si)']/1000)+1) *
        Si29_28_0
6
7   my_data['log_12C/13C'] = np.log10(my_data['12C/13C'])
8   my_data['log_14N/15N'] = np.log10(my_data['14N/15N'])
9   my_data['log_30Si/28Si'] = np.log10(my_data['30Si/28Si'])
10  my_data['log_29Si/28Si'] = np.log10(my_data['29Si/28Si'])
11
12  # Save to Excel
13  my_data.to_excel("sic_filtered_data.xlsx")
14
15  # Scvaling using StandardScaler() and RobustScaler()
16  X = my_data[['log_12C/13C','log_14N/15N','log_30Si/28Si','
        log_29Si/28Si']].values
17
18  scalers =[("Unscaled", X),
19          ("Standard Scaler",StandardScaler().fit_transform(X)),
20          ("Robust Scaler",RobustScaler().fit_transform(X))
21          ]
22
23  # Make pictures
24  fig = plt.figure(figsize=(15,8))
25
26  for ix, my_scaler in enumerate(scalers):
27      scaled_X = my_scaler[1]
28      ax = fig.add_subplot(2,3,ix+1)
29      ax.set_title(my_scaler[0])
30      ax.scatter(scaled_X[:,0], scaled_X[:,1],
31              marker='o', edgecolor='k', color='#db0f00',
```

```
32              alpha=0.6, s=40)
33      ax.set_xlabel(r'$log_{10}[^{12}C/^{13}C]$')
34      ax.set_ylabel(r'$log_{10}[^{14}N/^{15}N]$')
35
36      ax1 = fig.add_subplot(2,3,ix+4)
37      ax1.set_title(my_scaler[0])
38      ax1.scatter(scaled_X[:,2], scaled_X[:,3],
39              marker='o', edgecolor='k', color='#db0f00',
40              alpha=0.6, s=40)
41      ax1.set_xlabel(r'$log_{10}[^{30}Si/^{28}Si]$')
42      ax1.set_ylabel(r'$log_{10}[^{29}Si/^{28}Si]$')
43
44  fig.set_tight_layout(True)
```

Listing 3.10 Working example of data pre-processing (part 2)

Code listing 3.9 starts by importing all of the requisite libraries and methods (i.e., pandas, matplotlib, numpy, plus *StandardScaler* and *RobustScaler* from scikit-learn). The workflow starts at line 8, where we create a pandas DataFrame named *my_data*, importing the data set of SiC analyses from Excel™. All subsequent steps prepare *my_data* for processing by a ML algorithm.

Note that, in code listing 3.9,

Line 12   Limits the features to those of interest.
Line 19   Removes non-numerical data (i.e., Not a Number, or NaN).
Line 22   Removes all the rows labeled by "M" in the "PGD Type" column and characterized by large errors.
Line 27   Limits the data set to specific labels in the PGD-Type column (i.e., specific SiC classes such as X, N, AB, M, Y, and Z, in agreement with the current classification) (Stephan et al., 2021).
Line 35   Removes contaminated grains, that is, those characterized by an isotopic signature too similar to that of the Earth.

Then, in code listing 3.10,

Lines 2–5   Convert silica values from δ notation to isotopic ratios.
Lines 7–10   Apply a log transformation, consistent with the *alr* CoDA transformation.
Line 13   Save *my_data* to Excel™ to record the results of pre-processing before scaling.
Line 16   Define X, a four-feature numpy array in the shape accepted by most scikit-learn ML algorithms.
Line 18   Defines three scenarios: (1) unscaled data, (2) scaling with StandardScaler(), and (3) scaling with RobustScaler().
Lines 24–42   Perform the scaling (line 27) and show the diagrams in Fig. 3.4.

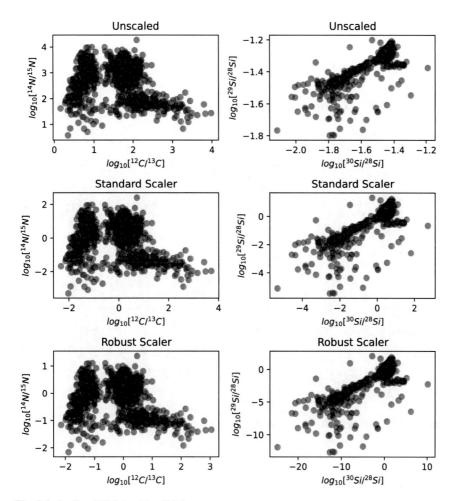

Fig. 3.4 Scaling SiC data with scikit-learn

Figure 3.4 shows the results of code listings 3.9 and 3.10. As expected, the application of various scalers and transformers does not change the data structure. However, it strongly affects the position and the spread of the features investigated. For example, the logarithm of $^{12}C/^{13}C$ ranges from 0 to 4 when unscaled, with a mean at about 1.7 (see also Fig. 3.3). Both the standard and the robust scalers center the data set on zero by using the mean and the median, respectively, but they produce different spreads because the robust scaler also accounts for the presence of outliers. For symmetric distributions in the absence of outliers, we expect similar results from the standard and robust scalers.

```
 1  import numpy as np
 2  import matplotlib.pyplot as plt
 3  from sklearn.preprocessing import StandardScaler
 4  from sklearn.mixture import GaussianMixture as GMM
 5
 6  my_colors = ['#AF41A5','#0A3A54','#0F7F8B','#BFD7EA','#F15C61',
 7              '#C82127','#ADADAD','#FFFFFF', '#EABD00']
 8
 9  scaler = StandardScaler().fit(X)
10  scaled_X = scaler.transform(X)
11
12  my_model = GMM(n_components = 9, random_state=(42)).fit(scaled_X)
13
14  Y = my_model.predict(scaled_X)
15
16  fig, ax = plt.subplots()
17
18  for my_group in np.unique(Y):
19      i = np.where(Y == my_group)
20      ax.scatter(scaled_X[i,0], scaled_X[i,1],
21              color=my_colors[my_group],
22              label=my_group + 1 , edgecolor='k', alpha=0.8)
23
24  ax.legend(title='Cluster')
25
26  ax.set_xlabel(r'$log_{10}[^{12}C/^{13}C]$')
27  ax.set_ylabel(r'$log_{10}[^{14}N/^{15}N]$')
28
29  fig.tight_layout()
```

Listing 3.11 Application of the *GaussianMixture*() algorithm to SiC data

3.4   Training a Model

Figure 3.5 shows a cheat sheet that guides us in selecting a model for the scikit-learn library.[9]

Scikit-learn works in the fields of both unsupervised learning (i.e., clustering and dimensionality reduction) and supervised learning (i.e., regression and classification). In supervised learning, examples of classification algorithms are the support vector classifier (see Sect. 7.9) and the K-nearest neighbors (see Sect. 7.10). In the field of regression, examples are the stochastic gradient descent (SGD), support vector (SVR), and ensemble regressors. Examples of unsupervised learning, if we consider dimensionality reduction, are locally linear embedding (LLE, see Sect. 4.3) and principal component analysis (PCA, see Sect. 4.2). For clustering, examples are K means, Gaussian mixture models (GMM, see Sect. 4.9), and spectral clustering.

---

[9] https://scikit-learn.org/stable/tutorial/machine_learning_map/.

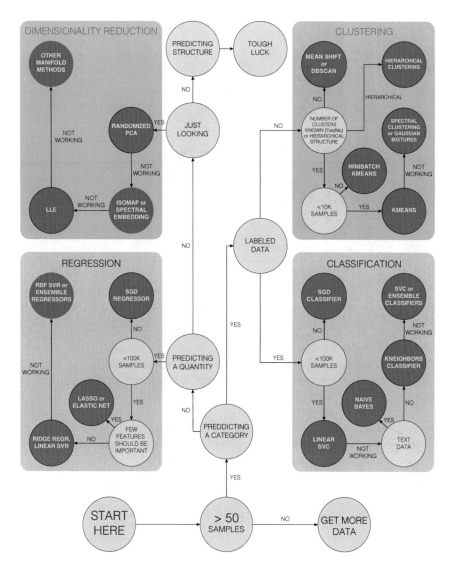

Fig. 3.5 Scikit-learn algorithm cheat sheet. Modified from the official documentation of scikit-learn

We discuss in detail the most popular ML algorithms in Chaps. 4 and 7, which deal with unsupervised and supervised learning, respectively.

I now present a simple example of training an unsupervised algorithm for SiC analyses that we use as a proxy for a scientific data set in the field of geochemistry and cosmochemistry science. Code listing 3.11 shows how to cluster SiC data by Gaussian mixtures (see Sect. 4.9) with the data previously pre-processed by code listings 3.9 and 3.10. The core of the training is at line 12, where I parameterized the

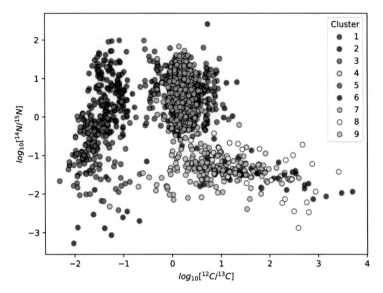

Fig. 3.6 Clustering produced by applying the *GaussianMixture*() algorithm to the SiC data (code listing 3.11)

GaussianMixture() algorithm (i.e., defining nine clusters and fixing the random state of the pseudo-random number generator to allow the reader to exactly reproduce my results).

Generally speaking, the *.fit*() method in scikit-learn launches the training of ML algorithms. Then, using the *.predict*() method, we get the results or we transfer the knowledge obtained to unknown data. Figure 3.6 shows the result of clustering by *GaussianMixture*() (see lines 16–29 of code listing 3.11).

3.5   Model Validation and Testing

The validation and testing of a model is the third fundamental step in ML, after pre-processing and training. They allow us to evaluate the "goodness" of a model.

3.5.1   Splitting the Investigated Data Set into Three Parts

The approach of model validation and testing by splitting the investigated data set into three parts is clearly described by Hastie et al. (2017) (Fig. 3.7): the best approach for model assessment in ML "is to randomly divide the data set into three parts: a training set, a validation set, and a test set. The training set is used to fit the

Fig. 3.7   Splitting the investigated data set into three parts

models; the validation set is used to estimate prediction error for model selection; the test set is used for assessment of the generalization error of the final chosen model."

```
1 from sklearn import preprocessing
2 from sklearn.model_selection import train_test_split
3
4 le = preprocessing.LabelEncoder()
5 le.fit(my_data['PGD Type'])
6 y = le.transform(my_data['PGD Type'])
7
8 X_train_valid, X_test, y_train_valid, y_test = train_test_split(
9         X, y, test_size=0.20)
10
11 X_train, X_valid, y_train, y_valid = train_test_split(
12         X, y, test_size=0.25)
```

Listing 3.12   Splitting the investigated data set into three parts in scikit-learn

Typically, we use the training data set to train a selection of candidate models, which could be different algorithms, a single algorithm tuned with different hyper-parameters (i.e., one or more variables that affect the behavior of an algorithm), or a combination of both. We then use the validation data set to evaluate candidate models and, based on the results, choose the best model. Finally, we check the selected model using the test data set. As an example, the *train_test_split*() method in scikit-learn randomly splits a data set into two parts (e.g., training plus validation and test sets). Again applying the *train_test_split*() method to the training plus validation set further divides it into the training and validation sets.

Note that the statements on lines 4–6 of code listing 3.12 simply convert the labels referring to a specific SiC Class (i.e., M, Y, Z, X, AB, and N) to an integer value ranging from 0 to 5. This approach facilitates the management of labels during the execution of supervised methods in the fields of regressions and classification.

3.5.2   Cross-Validation

The cross-validation (CV) procedure may be seen as an evolution of the static division of the investigated data set into three parts.

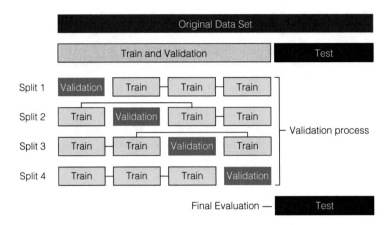

Fig. 3.8 Example of k-fold cross-validation

In the cross-validation procedure, the initial data set is split into two parts: the test set and the training plus validation sets. Then, in the most basic strategy of cross-validation (called k-fold CV), the joint training and validation set is split into k smaller batches (Fig. 3.8). The following steps consist of repeating the training and the validation for the candidate model as follows: (1) we use $k - 1$ folds as the training set; (2) the result of the training is validated against the remaining fold of the data; and (3) we repeat the procedure for the next split.

```
1 from sklearn import svm
2 from sklearn import preprocessing
3 from sklearn.model_selection import cross_validate
4
5 le = preprocessing.LabelEncoder()
6 le.fit(my_data['PGD Type'])
7 y = le.transform(my_data['PGD Type'])
8
9 my_model = svm.SVC(kernel='linear', C=1, random_state=42)
10
11 cv_results = cross_validate(my_model, scaled_X, y, cv=5,
12                            scoring='accuracy')
13
14 print(cv_results['test_score'])
15
16 '''
17 Output:
18 [0.98529412 0.97785978 0.9704797  0.98154982 0.95940959]
19 '''
```

Listing 3.13 Application of a linear support vector classifier to SiC data

```
1  from sklearn import svm
2  from sklearn import preprocessing
3  from sklearn.model_selection import GridSearchCV
4
5  le = preprocessing.LabelEncoder()
6  le.fit(my_data['PGD Type'])
7  y = le.transform(my_data['PGD Type'])
8
9  parameters = {'kernel':('linear', 'rbf'), 'C':[0.1, 1, 10]}
10 my_model = svm.SVC()
11
12 my_grid_search = GridSearchCV(my_model, parameters,
13                               cv = 4, scoring='accuracy')
14
15 my_grid_search.fit(scaled_X, y)
```

Listing 3.14 Model evaluation and selection by *k*-fold CV

The performance of the candidate model can be estimated by using the selected metrics and averaging the *k* results obtained. As an example, code listing 3.13 shows how to perform *k*-fold CV in scikit-learn using the *cross_validate*() method. After converting the five labels in the "PGD Type" columns (i.e., M, Y, Z, X, AB, N to a numeric index ranging from 0 to 5; see lines 5 to 7), we define a linear support vector classifier (see Sect. 7.9) characterized by a $C = 1$ hyperparameter (line 9). Finally, we perform the *k*-fold CV by dividing the data set into fivefold and using accuracy as a metric. As expected, we obtain five estimates for the accuracy, one for each split.

```
1  In [01]: my_grid_search.best_estimator_
2  Out[01]: SVC(C=10, kernel='linear')
3
4  In [02]: my_grid_search.best_score_
5  Out[02]: 0.9778761061946903
6
7  In [03]: my_grid_search.cv_results_
8  Out[03]:
9  {'mean_fit_time': array([0.00605977, 0.02105349, 0.00482285,
10             0.01113951, 0.00554657, 0.00662667]),
11  'std_fit_time': array([3.7539e-04, 6.0314e-04, 2.1346e-04,
12             7.0395e-04, 5.5384e-04, 3.1989e-05]),
13  'mean_score_time': array([0.00242817, 0.01987976, 0.00181627,
14             0.00979179, 0.00133586,0.00618142]),
15  'std_score_time': array([7.4277e-05, 1.6316e-03, 1.6929e-04,
16             2.7074e-04, 2.2063e-04, 6.4881e-04]),
17  'param_C': masked_array(data=[0.1, 0.1, 1, 1, 10, 10],
18             mask=[False, False, False, False, False, False],
19             fill_value='?', dtype=object),
20  'param_kernel': masked_array(data=['linear', 'rbf', 'linear',
21             'rbf', 'linear', 'rbf'],
22             mask=[False, False, False, False, False, False],
23             fill_value='?', dtype=object),
```

```
24   'params':    [{'C': 0.1, 'kernel': 'linear'},
25           {'C': 0.1, 'kernel': 'rbf'},
26           {'C': 1, 'kernel': 'linear'},
27           {'C': 1, 'kernel': 'rbf'},
28           {'C': 10, 'kernel': 'linear'},
29           {'C': 10, 'kernel': 'rbf'}],
30   'split0_test_score': array([0.92330383, 0.8879056 , 0.98230088,
31               0.91150442, 0.97935103, 0.97050147]),
32   'split1_test_score': array([0.9380531 , 0.88495575, 0.97935103,
33               0.92625369, 0.98525074, 0.97935103]),
34   'split2_test_score': array([0.92330383, 0.89380531, 0.97345133,
35               0.91740413, 0.97640118, 0.96460177]),
36   'split3_test_score': array([0.91740413, 0.88495575, 0.96755162,
37               0.90560472, 0.97050147, 0.96460177]),
38   'mean_test_score': array([0.92551622, 0.8879056 , 0.97566372,
39               0.91519174, 0.97787611, 0.96976401]),
40   'std_test_score': array([0.00762838, 0.00361282, 0.00566456,
41               0.00762838, 0.00531792, 0.0060364 ]),
42   'rank_test_score': array([4, 6, 2, 5, 1, 3], dtype=int32)}
```

Listing 3.15 How to get the results of *GridSearchCV()*

Using the *k*-fold cross-validation, *n* different candidate models can be evaluated by repeating *n* times the *k*-fold CV. As an example, the *GridSearchCV()* method in scikit-learn performs an exhaustive search (i.e., it evaluates all possible combinations of the proposed parameters) over a range of parameter values for a specific estimator (i.e., a ML algorithm). As an example, the method *GridSearchCV()* can be used to determine the best choice for the hyperparameters of a ML algorithm, such as the *C* parameter and the "kernel function" of a support vector machine (see Sect. 7.9). The code listing 3.14 shows in detail how to define the grid for the selected hyperparameters (line 9). On line 10, we define the model (i.e., a support vector classifier). On line 12, we define the grid search for our support vector classifier model, using the parameters defined on line 9, a fourfold cross-validation, and accuracy as a metric. Finally, on line 15 we physically perform the grid search for all combinations among the defined parameters. In detail, line 9 defines two kernel functions and three values for *C*. Therefore, the grid search performs six cross-validations and splits the *scaled_X* data set into four folds.

Code listing 3.15 shows how to get the results of a *GridSearchCV()*. More specifically, the *best_estimator_*, *best_score_*, and *cv_results_* attributes provide us with the optimal combination of hyperparameters, the best score, and a dictionary containing all the results, respectively.

3.5.3 Leave-One-Out Cross-Validation

The Leave-one-out (or LOO) cross-validation is a limiting case of the *k*-fold CV. When using the LOO approach, each training set is created by taking all the samples except one. The test set is then created by using the sample left out.

```
1  import numpy as np
2  from sklearn import svm
3  from sklearn.model_selection import LeaveOneOut
4  from sklearn.model_selection import cross_validate
5  import matplotlib.pyplot as plt
6
7  loo = LeaveOneOut()
8
9  my_model = svm.SVC(kernel='linear', C=1, random_state=42)
10
11 cv_results = cross_validate(my_model, scaled_X, y, cv=loo,
12                             scoring='accuracy')
13
14 fig, ax = plt.subplots()
15 my_x = [0,1]
16 my_height = [np.count_nonzero(cv_results['test_score'] == 0),
17              np.count_nonzero(cv_results['test_score'] == 1)]
18 my_bar = ax.bar(x = my_x, height=my_height, width=1,
19               color=['#F15C61', '#BFD7EA'],
20               tick_label=['wrongly classified', 'correcty
    classified'],
21               edgecolor='k')
22 ax.set_ylabel('occurrences')
23 ax.set_title('LOO cross validation n = {}'.format(len(scaled_X)))
24 ax.bar_label(my_bar)
25 ax.set_ylim(0,1600)
```

Listing 3.16 Leave-one-out cross-validation

In the LOO approach, the cross-validation typically covers all potential training sets (i.e., each sample of the investigated data set). Code listing 3.16 highlights how to perform a LOO cross-validation on the same study case used in code listing 3.13. Figure 3.9 shows the results of the LOO cross-validation of code listing 3.16. In the specific case under study, code listing 3.13 cross-validates 1356 models, each of which considers one of the investigated samples as the test data set, with all other samples serving for training.

3.5.4   Metrics

As you have probably noticed, the validation process is based on a metric. As an example, code listings 3.13, 3.13, and 3.16 specify *scoring='accuracy'*, which means that all examples given to this point use accuracy as a metric to quantify the "goodness" of a model. Note that a plethora of metrics exist that can potentially be used to validate a model. For example, Tables 3.3, 3.4, and 3.5 list the metrics that are available in scikit-learn for classification, regression, and clustering,

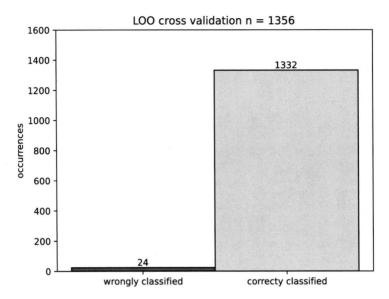

Fig. 3.9 Result of the LOO cross-validation (code listing 3.16)

respectively.[10] All the metrics reported in these tables follow the same convention: the goodness of the model increases as the value returned by the selected metric increases. In other words, higher values for a specific metric are better than lower values.

3.5.5 Overfitting and Underfitting

Over- and under-fitting should definitively be avoided when training a ML model. Over-fitting is when the trained models work suspiciously well in fitting the training set, whereas the performance with real-world data is poor (Shai & Shai, 2014). In other words, over-fitting occurs "when our hypothesis fits the training data *too well* (Shai & Shai, 2014)." Conversely, when our hypothesis is too simplistic (e.g., we try training a linear model to fit a nonlinear pattern; see Fig. 3.10) we have under-fitting, meaning a large approximation error (Shai & Shai, 2014).

---

[10] https://scikit-learn.org/stable/modules/model_evaluation.html.

Table 3.3 Metrics and scoring for the classification in scikit-learn

Method in metrics	Keywords	Description
.accuracy_score	'accuracy'	Accuracy classification score
.balanced_accuracy_score	'balanced_accuracy'	Compute the balanced accuracy
.top_k_accuracy_score	'top_k_accuracy'	Top-k Accuracy classification
.average_precision_score	'average_precision'	Compute the average precision
.brier_score_loss	'neg_brier_score'	Compute the Brier score loss
.precision_score	'precision' 'precision_micro' 'precision_macro' 'precision_weighted' 'precision_samples'	Compute the precision
.f1_score	'f1' 'f1_micro' 'f1_macro' 'f1_weighted' 'f1_samples'	Compute the F1 score
.recall_score	'recall' 'recall_micro' 'recall_macro' 'recall_weighted' 'recall_samples'	Compute the recall
.jaccard_score	'jaccard' 'jaccard_micro' 'jaccard_macro' 'jaccard_weighted' 'jaccard_samples'	Jaccard similarity coefficient
.roc_auc_score	'roc_auc' 'roc_auc_ovr' 'roc_auc_ovo' 'roc_auc_ovr_weighted' 'roc_auc_ovo_weighted'	Area Under the Receiver Operating Characteristic Curve (ROC AUC)

3.6 Model Deployment and Persistence

The deployment and persistence of a ML model is the last step of our workflow. Many options exist to ensure the persistence of a model, such as the use of pickles, joblib's pipelines, the Open Neural Network Exchange Format,[11] and the Predictive Model Markup Language[12] format.

---

[11] https://onnx.ai.

[12] https://dmg.org.

Table 3.4 Metrics and scoring for the regression in scikit-learn

Method in metrics	Keywords	Description
.explained_variance_score	'explained_variance'	Explained variance regression score
.max_error	'max_error'	Calculates the maximum residual error
.mean_absolute_error	'neg_mean_absolute_error'	Mean absolute error regression loss
.mean_squared_error	'neg_mean_squared_error'	Mean squared error regression loss
	'neg_root_mean_squared_error'	Root mean squared error regression loss
.mean_squared_log_error	'neg_mean_squared_log_error'	Mean squared logarithmic error regression loss
.median_absolute_error	'neg_median_absolute_error'	Median absolute error regression loss
.r2_score	'r2'	R^2-coefficient of determination score
.mean_poisson_deviance	'neg_mean_poisson_deviance'	Mean Poisson deviance regression loss
.mean_gamma_deviance	'neg_mean_gamma_deviance'	Mean Gamma deviance regression loss
.mean_absolute_percentage_error	'neg_mean_absolute_ percentage_error'	Mean absolute percentage error regression loss

As reported in the scikit-learn official documentation,[13] joblib's pipelines share some maintenance and security issues. For example, they assume the deployment of models in the same environment (i.e., the same library versions and Python core). Due to the above-mentioned issues, I suggest using the Open Neural Network Exchange Format or the Predictive Model Markup Language format to ensure the persistence of your ML model. These formats aim to improve model portability on different computing architectures and long-term archiving.

---

[13] https://scikit-learn.org/stable/model_persistence.html.

Table 3.5 Metrics and scoring for the clustering in scikit-learn

Method in metrics	Keywords	Description
.adjusted_mutual_info_score	'adjusted_mutual_info_score'	Adjusted mutual information between two clusterings
.adjusted_rand_score	'adjusted_rand_score'	Rand index adjusted for chance
.completeness_score	'completeness_score'	Completeness metric of a cluster labeling given a ground truth
.fowlkes_mallows_score	'fowlkes_mallows_score'	Measure the similarity of two clusterings of a set of points
.homogeneity_score	'homogeneity_score'	Homogeneity metric of a cluster labeling given a ground truth
.mutual_info_score	'mutual_info_score'	Mutual Information between two clusterings
.normalized_mutual _info_score	'normalized_mutual _info_score'	Normalized Mutual Information between two clusterings
.rand_score	'rand_score'	Rand index
.v_measure_score	'v_measure_score'	V-measure cluster labeling given a ground truth

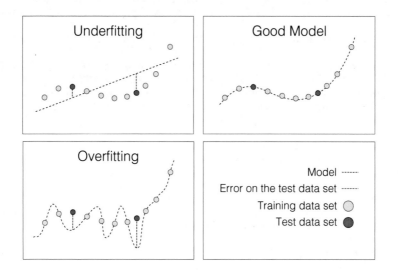

Fig. 3.10 Overfitting and underfitting

References

Aitchison, J. (1982). The statistical analysis of compositional data. *Journal of the Royal Statistical Society. Series B (Methodological)*, *44*(2), 139–177.

Aitchison, J. (1984). The statistical analysis of geochemical compositions. *Journal of the International Association for Mathematical Geology*, *16*(6), 531–564.

Aitchison, J., & Egozcue, J. J. (2005). Compositional data analysis: Where are we and where should we be heading? *Mathematical Geology*, *37*(7), 829–850. https://doi.org/10.1007/S11004-005-7383-7

Bestagini, P., Lipari, V., & Tubaro, S. (2017). A machine learning approach to facies classification using well logs. In *SEG Technical Program Expanded Abstracts* (pp. 2137–2142). https://doi.org/10.1190/SEGAM2017-17729805.1

Boujibar, A., Howell, S., Zhang, S., Hystad, G., Prabhu, A., Liu, N., Stephan, T., Narkar, S., Eleish, A., Morrison, S. M., Hazen, R. M., & Nittler, L. R. (2021). Cluster analysis of presolar silicon carbide grains: Evaluation of their classification and astrophysical implications. *The Astrophysical Journal. Letters*, *907*(2), L39. https://doi.org/10.3847/2041-8213/ABD102

Corlett, W. J., Aitchison, J., & Brown, J. A. C. (1957). The lognormal distribution, with special reference to its uses in economics. *Applied Statistics*, *6*(3), 228. https://doi.org/10.2307/2985613

De Mauro, A., Greco, M., & Grimaldi, M. (2016). A formal definition of Big Data based on its essential features. *Library Review*, *65*(3), 122–135. https://doi.org/10.1108/LR-06-2015-0061/FULL/XML

Egozcue, J. J., & Pawlowsky-Glahn, V. (2005). Groups of parts and their balances in compositional data analysis. *Mathematical Geology*, *37*(7), 795–828. https://doi.org/10.1007/S11004-005-7381-9

Hastie, T., Tibshirani, R., & Friedman, J. (2017). *The elements of statistical learning* (2nd ed.). Springer.

Limpert, E., Stahel, W. A., & Abbt, M. (2001). Log-normal distributions across the sciences: Keys and clues. https://doi.org/10.1641/0006-3568(2001)051[0341:LNDATS]2.0.CO;2

Maharana, K., Mondal, S., & Nemade, B. (2022). A review: Data pre-processing and data augmentation techniques. In *Global Transitions Proceedings*. https://doi.org/10.1016/J.GLTP.2022.04.020

Panda, D. K., Lu, X., & Shankar, D. (2022). *High-performance big data computing*. MIT Press.

Pawlowsky-Glahn, V., & Buccianti, A. (2011). *Compositional data analysis*. Wiley Online Library.

Petrelli, M. (2021). *Introduction to Python in earth science data analysis*. Springer International Publishing. https://doi.org/10.1007/978-3-030-78055-5

Pietsch, W. (2021). *Big Data*. Cambridge University Press. https://doi.org/10.1017/9781108588676

Razum, I., Ilijanić, N., Petrelli, M., Pawlowsky-Glahn, V., Miko, S., Moska, P., & Giaccio, B. (2023). Statistically coherent approach involving log-ratio transformation of geochemical data enabled tephra correlations of two late Pleistocene tephra from the eastern Adriatic shelf. *Quaternary Geochronology*, *74*, 101416. https://doi.org/10.1016/J.QUAGEO.2022.101416

Shai, S.-S., & Shai, B.-D. (2014). *Understanding machine learning: From theory to algorithms*. Cambridge University Press.

Stephan, T., Bose, M., Boujibar, A., Davis, A. M., Gyngard, F., Hoppe, P., Hynes, K. M., Liu, N., Nittler, L. R., Ogliore, R. C., & Trappitsch, R. (2021). The Presolar Grain Database for silicon carbide—grain type assignments (abstract). In *Lunar Planetary Science* (vol. *52*, p. 2358).

van den Boogaart, K. G., & Tolosana-Delgado, R. (2013). *Analyzing compositional data with R*. Springer Berlin Heidelberg. https://doi.org/10.1007/978-3-642-36809-7/COVER

Zhang, Z. (2016). Missing data imputation: focusing on single imputation. *Annals of Translational Medicine*, *4*(1), 9. https://doi.org/10.3978/J.ISSN.2305-5839.2015.12.38

Part II
Unsupervised Learning

Chapter 4
Unsupervised Machine Learning Methods

4.1 Unsupervised Algorithms

As introduced in Chap. 1, the unsupervised learning process acts on unlabeled data and attempts to extract significant patterns from the investigated data set. In the present chapter, I gently introduce the unsupervised algorithms for dimensionality reduction and clustering reported in Fig. 3.5. Finally, I provide some specific references to allow readers to delve deeper into the mathematics that governs these ML methods. In detail, I start by describing the algorithms for dimensionality reduction, which include the principal component analysis and methods based on manifold learning. I then describe clustering methods, such as hierarchical clustering, DBSCAN, mean shift, K means, spectral clustering, and Gaussian mixtures models.

4.2 Principal Component Analysis

Principal component analysis (PCA) is a multivariate statistical method that extracts relevant information from a data set and represents it in a lower-dimensional space (Jollife & Cadima, 2016). It strives to increase the interpretability of a data set by reducing the dimensionality of the problem while at the same time minimizing information loss (Jollife & Cadima, 2016). In detail, it creates new uncorrelated variables (i.e., through a linear combination of the original variables), called "principal components," that maximize variance (Jollife & Cadima, 2016).

M. Petrelli, *Machine Learning for Earth Sciences*, Springer Textbooks
in Earth Sciences, Geography and Environment,
https://doi.org/10.1007/978-3-031-35114-3_4

Mathematically, PCA is an eigenvalue-eigenvector problem (Jollife & Cadima, 2016). Consider a d-dimensional sample set $X = \{\mathbf{x}_1, \mathbf{x}_2, \mathbf{x}_j, \ldots, \mathbf{x}_p\}$ made of n observations on p numerical variables. The sample set X is equivalent to an $n \times p$ data matrix $\mathbf{X}$, whose jth column is the vector $\mathbf{x}_j$ of observations on the jth variable (Jollife & Cadima, 2016). We look for a linear combination of the columns of matrix $\mathbf{X}$ with maximum variance (Jollife & Cadima, 2016). Such linear combinations are given by

$$\sum_{j=1}^{p} a_j \mathbf{x}_j = \mathbf{X}\mathbf{a}, \tag{4.1}$$

where $\mathbf{a} = \{a_1, a_2, \ldots, a_p\}$ is a vector of constants (Jollife & Cadima, 2016). The variance of any linear combination defined by Eq. (4.1) is given by Jollife and Cadima (2016)

$$var(\mathbf{X}\mathbf{a}) = \mathbf{a}^T \mathbf{S}\mathbf{a}, \tag{4.2}$$

where $\mathbf{S}$ is the sample covariance matrix associated with the data set (Jollife & Cadima, 2016).

The solution to the problem (i.e., identifying the linear combination with maximum variance) consists of finding a d-dimensional vector $\mathbf{a}$ that maximizes the quadratic form $\mathbf{a}^T \mathbf{S}\mathbf{a}$ (Jollife & Cadima, 2016). To obtain a defined solution, the most common restriction assumes working with unit-norm vectors (i.e., requiring $\mathbf{a}^T \mathbf{a} = 1$). Now the problem is equivalent to maximizing the relation (Jollife & Cadima, 2016)

$$\mathbf{a}^T \mathbf{S}\mathbf{a} - \lambda \left(\mathbf{a}^T \mathbf{a} - 1 \right). \tag{4.3}$$

After differentiating with respect to the vector $\mathbf{a}$ and equating to the null vector, we have (Jollife & Cadima, 2016)

$$\mathbf{S}\mathbf{a} = \lambda \mathbf{a}. \tag{4.4}$$

In Eq. (4.4), $\mathbf{a}$ is a unit-norm eigenvector and λ is the corresponding eigenvalue of $\mathbf{S}$ (Jollife & Cadima, 2016). The full set of eigenvectors of $\mathbf{S}$ are the solutions to the problem of obtaining up to d new linear combinations $\mathbf{X}\mathbf{a}_k = \sum_{j=1}^{d} a_{jk}\mathbf{x}_j$, which successively maximize variance subject to noncorrelation with previous linear combinations (Jolliffe, 2002; Jollife & Cadima, 2016).

4.3   Manifold Learning

The main idea behind manifold learning methods is that, although natural data sets are often depicted in very-high-dimensional spaces, they can be described in lower dimensions because the processes generating the data are often characterized by few degrees of freedom (Zheng & Xue, 2009). From the mathematical point of view, manifold learning methods try to model the data as "lying on or near a low-dimensional manifold embedded in a higher-dimensional space" (Zheng & Xue, 2009). In the following, I introduce the basic concepts of manifold learning, but I strongly encourage you to go deeper into the details if you plan to use these techniques in your research (Zheng & Xue, 2009).

Manifold A d-dimensional manifold $\mathbb{M}$ is a topological space that is locally homeomorphic with respect to $\mathbb{R}^d$.

Homomorphism A map from one algebraic structure to another of the same type that preserves all the relevant structures.

Embedding An embedding of a manifold $\mathbb{M}$ into $\mathbb{R}^d$ is a smooth homeomorphism from $\mathbb{M}$ to a subset of $\mathbb{R}^d$.

4.3.1 Isometric Feature Mapping

The Isometric feature mapping (Isomap) is an ML algorithm that is "capable of discovering the nonlinear degrees of freedom that underlie complex natural observations" (Tenenbaum et al., 2000). It consists of three main steps: (1) construct a neighborhood graph, (2) compute the shortest paths, and (3) construct a d-dimensional embedding (Tenenbaum et al., 2000). In practice, Isomap searches for a lower-dimensional embedding while maintaining geodesic distances between all points. In scikit-learn, the method *Isomap()* performs the Isometric feature mapping.

4.3.2 Locally Linear Embedding

Locally linear embedding (LLE) (Roweis & Saul, 2000) is a ML algorithm that "computes low-dimensional, neighborhood-preserving embeddings of high-dimensional inputs" (Roweis & Saul, 2000). In practice, LLE maps the inputs onto a single global coordinate system of lower dimensionality (Roweis & Saul, 2000). Also, its optimizations do not involve local minima (Roweis & Saul, 2000). In other words, LLE searches for a lower-dimensional projection of the data while preserving distances within local neighborhoods. In scikit-learn, LLE is implemented in the method *LocallyLinearEmbedding()*.

4.3.3 Laplacian Eigenmaps

A Laplacian eigenmap (Belkin & Niyogi, 2003) first develops a graph incorporating neighborhood information starting from a data set in $\mathbb{R}^d$ and then uses the Laplacian to compute a low-dimensional representation. Practically, Laplacian eigenmaps consist of three main steps: (1) constructing the adjacency graph, (2) choosing the weights, and (3) computing the eigenmaps.

4.3.4 Hessian Eigenmaps

Hessian eigenmaps (Donoho & Grimes, 2003) are similar to Laplacian eigenmaps but replace the Laplacian operator with the Hessian operator. The main difference between Laplacian and Hessian eigenmaps relies on the capability of Hessian eigenmaps to overcome the 'convexity limitation' of Laplacian eigenmaps (Zheng & Xue, 2009). In scikit-learn, Hessian eigenmaps can be performed with the *LocallyLinearEmbedding()*, i.e., the same that we use for the LLE, but specifying *method* = *'hessian'*.

4.4 Hierarchical Clustering

Hierarchical clustering algorithms (Johnson, 1967) build a hierarchical representation of the data set structure where clusters at each level of the hierarchy are assembled by merging or splitting clusters at the next lower or upper level, respectively (Johnson, 1967; Hastie et al., 2017). Two main paradigms of hierarchical clustering exist: agglomerative (i.e., bottom-up) and divisive (i.e., top-down). Agglomerative strategies start from the bottom where every observation forms a cluster (Johnson, 1967; Hastie et al., 2017). Next, at each successive level, the algorithm recursively merges a selected pair of clusters into a single cluster. The criterion for merging (i.e., linkage) is based on specific metrics (Johnson, 1967; Hastie et al., 2017).

In contrast, the divisive approach starts from a single cluster containing all observations and, at each subsequent level, recursively splits one of the existent clusters into two new clusters using a dissimilarity metric (Johnson, 1967; Hastie et al., 2017). In scikit-learn, the method *AgglomerativeClustering()* performs the agglomerative hierarchical clustering using a bottom-up approach. The linkage criterion is based on the concept of dissimilarity. To understand this concept, consider two sets of observations; clusters G and H. Hierarchical clustering estimates the dissimilarity $d(G, H)$ between G and H on the set of pairwise-observation dissimilarities d_{ij}, where member i of the pair is in G and member j

Table 4.1 Linkage options in *AgglomerativeClustering*()

Parameter	Equation	Note
linkage='single'	$d_{sl}(G, H) = \min\limits_{\substack{i \in G \\ j \in H}} d_{ij}$	Uses the minimum of the distances between all observations of the two sets
linkage='complete'	$d_{cl}(G, H) = \max\limits_{\substack{i \in G \\ j \in H}} d_{ij}$	Uses the maximum distance between all observations of the two sets
linkage='average'	$d_{ga}(G, H) = \dfrac{1}{n_g n_h} \sum\limits_{i \in G} \sum\limits_{j \in H} d_{ij}$	Uses the average of the distances of each observation of the two sets

is in H (Hastie et al., 2017). Using *AgglomerativeClustering*(), the linkage criterion could be single, complete, group average, or Ward (Table 4.1).

Finally, Ward's linkage criterion (the default in scikit-learn) states that the distance between two clusters G and H is how much the sum of squares increases when they are merged:

$$\Delta(G, H) = \frac{|G| \, |H|}{|G| + |H|} \, \|m_G + m_H\|^2, \tag{4.5}$$

where Δ is the "merging cost" of combining clusters G and H. Also, m, $|G|$ and $|H|$ are the center of clusters and the cardinal of G and H, respectively.

The dissimilarities d_{ij} can be estimated by using different metrics. Using the method *AgglomerativeClustering*(), they can be "Euclidean" or "Manhattan," among others. For Ward linkage, the only metric accepted is "Euclidean" [see Eq. (4.5)].

4.5   Density-Based Spatial Clustering of Applications with Noise

The algorithm density-based spatial clustering of applications with noise (DBSCAN) relies on a "density-based notion of clusters which is designed to discover clusters of arbitrary shape" (Ester et al., 1996). Topologically, DBSCAN identifies a core sample if there exists a pre-defined minimum number of other (i.e., neighbors of the core sample) within a distance of ϵ (Ester et al., 1996). A cluster is a set of core samples plus their neighbors. Any sample that is neither a core sample nor a neighbor (i.e., it is at least a distance ϵ from any core sample) is marked as an outlier (Ester et al., 1996). Note that DBSCAN does not require the number of clusters to be specified.

4.6 Mean Shift

The mean shift algorithm is a nonparametric technique for clustering analysis (Comaniciu & Meer, 2002); it estimates the kernel density in the investigated d-dimensional feature space (Derpanis, 2005). As a result, the kernel density estimation defines an empirical probability density function where "dense regions" identify local maxima (i.e., modes) of the underlying distribution (Derpanis, 2005). Finally, the mean shift algorithm performs a gradient ascent (i.e., it searches until convergence for these maxima in the empirical probability density function) (Derpanis, 2005). In detail, the mean shift procedure for a given observation $\mathbf{x}_i$ is as follows (Derpanis, 2005; Comaniciu & Meer, 2002):

1. Compute the mean shift vector $\mathbf{m}(\mathbf{x}_i^t)$ at the step t;
2. Translate the density-estimation window: $\mathbf{x}_i^{t+1} = \mathbf{x}_i^t + m(\mathbf{x}_i^t)$;
3. Iterate steps 1 and 2 until convergence.

The mean shift vector is defined as follows [Eq. (17) in Comaniciu and Meer (2002)]:

$$\mathbf{m}(\mathbf{x}_i) = \left[\frac{\sum_{i=1}^{n} \mathbf{x}_i g\left(\left\|\frac{\mathbf{x}-\mathbf{x}_i}{h}\right\|^2\right)}{\sum_{i=1}^{n} g\left(\left\|\frac{\mathbf{x}-\mathbf{x}_i}{h}\right\|^2\right)} - \mathbf{x} \right], \tag{4.6}$$

where the function $g(x)$ is the derivative of the selected kernel estimator and h (i.e., the bandwidth parameter) defines the radius of the kernel (Comaniciu & Meer, 2002).

In scikit-learn, the *MeanShift*() method uses a flat kernel to perform mean shift clustering. Note that the default scikit-learn parametrization of the mean shift algorithm automatically sets the number of clusters and the optimal h (i.e., the bandwidth). However, h can be manually adjusted by using the *bandwidth* parameter.

4.7 *K*-Means

The K-means is a clustering technique that seeks to minimize the average squared distance between points in the same cluster (Arthur & Vassilvitskii, 2007). Note that the K-means algorithm requires the number of clusters to be specified. Mathematically, the K-means algorithm can be expressed as follows: given an integer k and a set of n data points in $\mathbb{R}^d$, the goal is to choose k centers to minimize the total squared distance between each point and its closest center (i.e., the inertia ϕ) (Arthur & Vassilvitskii, 2007):

$$\phi = \sum_{\mathbf{x} \in X} \min_{\mathbf{c} \in C} \|\mathbf{x} - \mathbf{c}\|^2 . \tag{4.7}$$

Usually, the K-means implementation (e.g., in scikit-learn) refers to the solution of the problem proposed by Lloyd (1982). In detail, the algorithm proposed by Arthur and Vassilvitskii (2007) consists of four steps:

1. Arbitrarily choose an initial k centers $C = \{\mathbf{c}_1, \mathbf{c}_2, \ldots, \mathbf{c}_k, \}$;
2. For each $i \in \{1, \ldots, k\}$, set the cluster Y_i to be the set of points in X that are closer to $\mathbf{c}_i$;
3. Define new centroids $\mathbf{c}_i$ by averaging all the samples assigned to each previous centroid;
4. Repeat steps 2 and 3 until C no longer changes significantly.

In scikit-learn, the method *KMeans()* implements K-means clustering. Also, *MiniBatchKMeans()* modifies the K-means algorithm by using minibatches to save computation time.

4.8 Spectral Clustering

Spectral clustering (Von Luxburg, 2007) is a ML technique that combines clustering with dimensionality reduction (Sugiyama, 2015). In detail, spectral clustering uses a kernel function to transform samples into a feature space and then applies a locality-preserving projection to reduce the dimensionality (see Fig. 4.1). Note that a locality-preserving projection in the feature space is equivalent to the Laplacian eigenmap manifold method described in Sect. 4.3.3 (Sugiyama, 2015). In practice, spectral clustering performs a low-dimensional embedding low-dimensional embedding of the similarity (or affinity) matrix between samples (Von Luxburg, 2007). Finally, spectral clustering uses a clustering method (e.g., K means) to obtain cluster labels (Sugiyama, 2015; Von Luxburg, 2007).

In scikit-learn the method *SpectralClustering()* applies spectral clustering. Note that *SpectralClustering()* requires the number of clusters to be specified in advance.

Fig. 4.1 Locality-preserving
projection. The projection
tries to maintain the cluster
structure when reducing the
dimensionality of the
problem. Modified from
Sugiyama (2015)

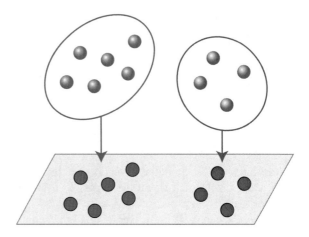

4.9   Gaussian Mixture Models

Gaussian mixture models (GMMs) try to reconstruct the probability density function
that underlies the investigated data set as generated by a mixture of a finite number
of Gaussian distributions with unknown parameters (McLachlan & Peel, 2000).

To understand how GMMs work, consider a d-dimensional (i.e., characterized
by d variables or features) sample set $X = \{\mathbf{x}_1, \mathbf{x}_2, \ldots, \mathbf{x}_n\}$ of independent and
identically distributed observations (McLachlan & Peel, 2000). Finite mixtures
models (FMMs) assume that the observations $\mathbf{x} \in X$ derive from a probability
density function described by a mixture of g components (McLachlan & Peel, 2000;
Scrucca et al., 2016):

$$f(\mathbf{x}, \psi) = \sum_{i=1}^{g} \pi_i f_i(\mathbf{x}, \boldsymbol{\theta}_i), \qquad (4.8)$$

where g and $\psi = \{\pi_1, \ldots, \pi_{g-1}, \boldsymbol{\theta}_1, \ldots, \boldsymbol{\theta}_g\}$ are the number of mixture components and the parameters of the model, respectively (Scrucca et al., 2016).
Also, $f_i(\mathbf{x}, \boldsymbol{\theta}_i)$ is the ith component density for the sample observation $\mathbf{x}$ and
is parametrized by the vector $\boldsymbol{\theta}_i$. Finally, $\{\pi_1, \ldots, \pi_{g-1}\}$ are the mixing weights
(Scrucca et al., 2016).

In many applications, the component densities $f_i(\mathbf{x}, \boldsymbol{\theta}_i)$ are assumed to belong
to the same parametric family (McLachlan & Peel, 2000). In some applications,
the component densities are taken to be different. The implementation of a finite
Gaussian mixtures model assumes $f_i(\mathbf{x}, \boldsymbol{\theta}_i)$ as a multivariate normal, a fixed G, and
consists of estimating the model parameters ψ (McLachlan & Peel, 2000).

In scikit-learn, the methods *GaussianMixture*() and *BayesianGaussianMixture*()
implement the finite Gaussian mixture model based on expectation-maximization
(EM) (Dempster et al., 1977) and variational Bayesian inference (Hastie et al.,
2017; Blei & Jordan, 2006), respectively. Variational Bayesian inference is similar

to expectation maximization, although the former adds a regularization step by integrating information from integrating information from prior distributions (Hastie et al., 2017; Blei & Jordan, 2006). The aim is to avoid pathological special cases, which often appear in expectation-maximization solutions (Blei & Jordan, 2006).

References

Arthur, D., & Vassilvitskii, S. (2007). K-Means++: The advantages of careful seeding. In *Proceedings of the Eighteenth Annual ACM-SIAM Symposium on Discrete Algorithms* (pp. 1027–1035).

Belkin, M., & Niyogi, P. (2003). Laplacian eigenmaps for dimensionality reduction and data representation. *Neural Computation, 15*(6), 1373–1396.

Blei, D. M., & Jordan, M. I. (2006). Variational inference for Dirichlet process mixtures. *Bayesian Analysis, 1*(1), 121–143. https://doi.org/10.1214/06-BA104

Comaniciu, D., & Meer, P. (2002). Mean shift: A robust approach toward feature space analysis. *IEEE Transactions on Pattern Analysis and Machine Intelligence, 24*(5), 603–619.

Dempster, A. P., Laird, N. M., & Rubin, D. B. (1977). Maximum likelihood from incomplete data via the EM algorithm. *Journal of the Royal Statistical Society: Series B (Methodological), 39*(1), 1–22. https://doi.org/10.1111/J.2517-6161.1977.TB01600.X

Derpanis, K. G. (2005). Mean shift clustering. In *Lecture Notes* (vol. 32).

Donoho, D. L., & Grimes, C. (2003). Hessian eigenmaps: Locally linear embedding techniques for high-dimensional data. *Proceedings of the National Academy of Sciences, 100*(10), 5591–5596.

Ester, M., Kriegel, H.-P., Sander, J., Xu, X., et al. (1996). A density-based algorithm for discovering clusters in large spatial databases with noise. In *kdd* (Vol. 96, No. 34, pp. 226–231).

Hastie, T., Tibshirani, R., & Friedman, J. (2017). *The elements of statistical learning* (2nd ed.). Springer.

Johnson, S. C. (1967). Hierarchical clustering schemes. *Psychometrika, 32*(3), 241–254.

Jolliffe, I. T. (2002). *Principal component analysis*. Springer-Verlag. https://doi.org/10.1007/B98835

Jollife, I. T., & Cadima, J. (2016). Principal component analysis: a review and recent developments. *Philosophical Transactions of the Royal Society A: Mathematical, Physical and Engineering Sciences, 374*(2065). https://doi.org/10.1098/RSTA.2015.0202

Lloyd, S. (1982). Least squares quantization in PCM. *IEEE Transactions on Information Theory, 28*(2), 129–137. https://doi.org/10.1109/TIT.1982.1056489

McLachlan, G. J., & Peel, D. (2000). *Finite mixture models*. Wiley.

Roweis, S. T., & Saul, L. K. (2000). Nonlinear dimensionality reduction by locally linear embedding. *Science, 290*(5500), 2323–2326. https://doi.org/10.1126/SCIENCE.290.5500.2323

Scrucca, L., Fop, M., Murphy, T. B., & Raftery, A. E. (2016). mclust 5: Clustering, classification and density estimation using gaussian finite mixture models. *The R Journal, 8*(1), 289–317. https://doi.org/10.32614/RJ-2016-021

Sugiyama, M. (2015). *Introduction to statistical machine learning*. Elsevier Inc. https://doi.org/10.1016/C2014-0-01992-2

Tenenbaum, J. B., De Silva, V., & Langford, J. C. (2000). A global geometric framework for nonlinear dimensionality reduction. *Science, 290*(5500), 2319–2323. https://doi.org/10.1126/SCIENCE.290.5500.2319

Von Luxburg, U. (2007). A tutorial on spectral clustering. *Statistics and Computing, 17*(4), 395–416.

Zheng, N., & Xue, J. (2009). Manifold Learning. In *Statistical learning and pattern analysis for image and video processing* (pp. 87–119). London: Springer. https://doi.org/10.1007/978-1-84882-312-94

Chapter 5
Clustering and Dimensionality Reduction in Petrology

5.1 Unveil the Chemical Record of a Volcanic Eruption

Unsupervised machine learning methods can help us decode the chemical record stored in the crystal cargo of a single eruption or multiple volcanic events (Caricchi et al., 2020b; Boschetty et al., 2022; Musu et al., 2023). This record often includes the major element's chemical composition (i.e., multivariate compositional data) of different crystal phases such as olivine, clinopyroxene, orthopyroxene, amphibole, plagioclase, garnet, and quartz (Boschetty et al., 2022; Aitchison & Egozcue, 2005; Aitchison, 1982, 1984). Each of these phases provides clues to unravel the complex dynamics of a volcanic plumbing system (Ubide et al., 2021) and its evolution (Costa et al., 2020; Petrelli & Zellmer, 2020).

During the crystallization process (Fig. 5.1), minerals grow and adapt their textural aspect and chemistry to the melt compositions and the thermodynamic conditions of the magmatic system (Ubide et al., 2021). For example, concentric chemical zones from the core to the rim of a crystal reflect the sequential changes over time imposed by the magmatic system (Fig. 5.1). Moderate-to-rapid growths at intermediate-to-high degrees of undercooling ($\Delta T = T_{\text{liquidus}} - T_{\text{crystallisation}}$) may result in sector zoning in euhedral crystals or skeletal to dendritic textures (Fig. 5.1). In addition, diffusive re-equilibration of compositional gradients can further modify chemical patterns in crystals (Costa et al., 2020; Petrelli & Zellmer, 2020).

At shallow crustal levels (Fig. 5.1), pre- and syn-eruptive dynamics include a complex range of processes, including magma fractionation, recharge, mixing, assimilation, and degassing (Ubide et al., 2021). Interrogating the crystal cargo of an eruption provides us with the requisite information to unravel the complex

© The Author(s), under exclusive license to Springer Nature Switzerland AG 2023
M. Petrelli, *Machine Learning for Earth Sciences*, Springer Textbooks in Earth Sciences, Geography and Environment, https://doi.org/10.1007/978-3-031-35114-3_5

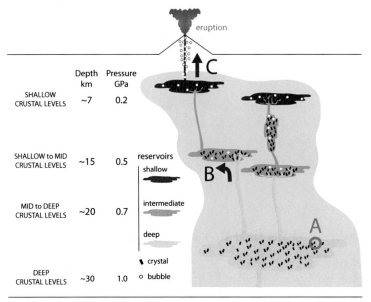

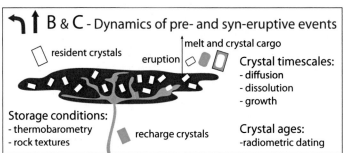

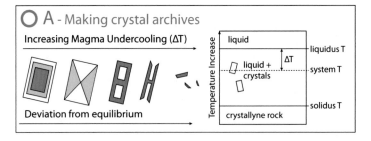

Fig. 5.1 Architecture of a volcanic plumbing system and related pre- and syn-eruptive dynamics. Modified from Petrelli and Zellmer (2020) and Ubide et al. (2021)

dynamics of a volcanic plumbing system before and during eruption (Ubide et al., 2021).

In this chapter, I focus on the data set reported by Musu et al. (2023), which consists of clinopyroxene analyses (cpx) erupted by the South-East Crater of Mt. Etna during the sequence of lava fountains that occurred between February and April of 2021 (Musu et al., 2023).

Musu et al. (2023) focused on cpx analyses because (1) cpx is typically found in mafic to intermediate magmas, (2) cpx crystallizes over a wide range of temperatures T and pressures P, and (3) cpx chemistry depends on magma composition, water content, pressure, and temperature (Musu et al., 2023), which make cpx a robust thermobarometer (Putirka, 2008; Petrelli et al., 2020; Jorgenson et al., 2022; Higgins et al., 2021) and a fine recorder of the chemical evolution of magmatic systems (Ubide & Kamber, 2018; Caricchi et al., 2020b; Boschetty et al., 2022).

5.2 Geological Setting

Mt. Etna is in eastern Sicily on the southern tip of the Italian peninsula (Fig. 5.2) and is the largest active volcano in Europe (Branca & Del Carlo, 2004) and one of the most active volcanoes in the world (Cappello et al., 2013; Corsaro & Miraglia, 2022).

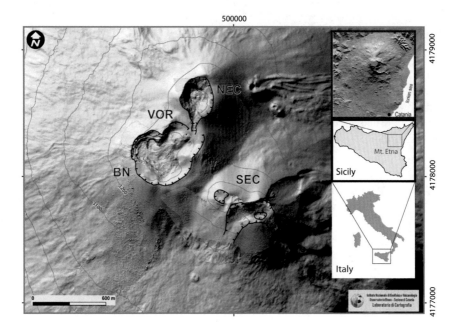

Fig. 5.2 Mt. Etna volcano. Modified from Musu et al. (2023)

The Mt. Etna volcano exhibits different eruptive behaviors, from effusive to explosive, including strombolian and violent lava-fountaining occurrences (Branca & Del Carlo, 2004; Ferlito et al., 2014; Corsaro & Miraglia, 2022). Eruptions come from summit craters and fissure vents along its flanks (Musu et al., 2023; Branca & Del Carlo, 2004; Di Renzo et al., 2019). The summit area consists of four active vents: Voragine (VOR), Bocca Nuova (BN), North-East Crater (NEC), and South-East Crater (SEC). Of these, the SEC is the youngest and most active vent (Andronico & Corsaro, 2011; Di Renzo et al., 2019; Corsaro & Miraglia, 2022).

A cyclical eruptive sequence started at the SEC on December 13, 2020 and generated over 60 paroxysms; in other words, "particularly violent eruptions of the volcano, which is the most dangerous and tense stage of this eruptive cycle, at which the whole cavity of the crater is opened" (Paffengoltz, 1978).

5.3 The Investigated Data Set

The data set contains major-element chemical analyses collected along rim-to-core transects on clinopyroxenes with a point spacing of $2\,\mu$m (Musu et al., 2023). A total of 1250 analyses were acquired (Musu et al., 2023) by using a JEOL 8200 Superprobe at the University of Geneva and a JEOL JXA-8530F at the University of Lausanne (Musu et al., 2023). Clinopyroxene samples belong to lapilli collected from the lava-fountain deposits of February 16, 19, and 28 and March 2 and 10, 2021.

5.4 Data Pre-processing

Code listings 5.1 and 5.2 reveal our data pre-processing strategy, including the final step of data visualization. The strategy consists of first cleaning the data and then transforming it for compositional data analysis (CoDA; cf. Sect. 3.3.6) and "robust" normalization. Finally, the resulting CoDA-transformed and -scaled data are visualized.

5.4.1 Data Cleaning

Code listing 5.1 is mainly a preliminary data-cleaning procedure. In detail, the function *calc_cations_on_oxygen_basis*() (lines 4–29) calculates the number of cations deriving from a specific chemical analysis based on a fixed number of oxygens in the chemical formula of a specific crystal phase. We are dealing with clinopyroxene analyses, so the base chemical formula contains six oxygens and four cations (line 36). Also, we define a tolerance of 0.06, which means that we discard all analyses that return less than 3.94 or more than 4.06 cations in the formula. We are mainly discarding bad chemical analyses (e.g., those affected by contamination, melt contamination, or additional issues). If you do not understand this step, please refer to an introductory text on mineralogy for further details (Okrusch & Frimmel, 2020). Another test for anhydrous crystal phases is to check for closure (i.e., verify that the sum of the oxides is close to 100 wt. %; lines 32 and 33).

```
1  import numpy as np
2  import pandas as pd
3
4  def calc_cations_on_oxygen_basis(myData0, my_ph, my_el, n_ox):
5      Weights = {
6          'SiO2': [60.0843,1.0,2.0], 'TiO2':[79.8788,1.0,2.0],
7          'Al2O3': [101.961,2.0,3.0],'FeO':[71.8464,1.0,1.0],
8          'MgO':[40.3044,1.0,1.0], 'MnO':[70.9375,1.0,1.0],
9          'CaO':[56.0774,1.0,1.0], 'Na2O':[61.9789,2.0,1.0],
10         'K2O':[94.196,2.0,1.0], 'Cr2O3':[151.9982,2.0,3.0],
11         'P2O5':[141.937,2.0,5.0], 'H2O':[18.01388,2.0,1.0]}
12     myData = myData0.copy()
13     myData = myData.add_prefix(my_ph + '_')
14     for el in my_el: # Cation mole proportions
15         myData[el + '_cat_mol_prop'] = myData[my_ph +
16             '_' + el] * Weights[el][1] / Weights[el][0]
17     for el in my_el:  # Oxygen mole proportions
18         myData[el + '_oxy_mol_prop'] = myData[my_ph +
19             '_' + el] * Weights[el][2] / Weights[el][0]
20     totals = np.zeros(len(myData.index)) # Ox mole prop tot
21     for el in my_el:
22         totals += myData[el + '_oxy_mol_prop']
23     myData['tot_oxy_prop'] = totals
24     totals = np.zeros(len(myData.index)) # totcations
25     for el in my_el:
26         myData[el + '_num_cat'] = n_ox * myData[el +
27             '_cat_mol_prop'] / myData['tot_oxy_prop']
28         totals += myData[el + '_num_cat']
29     return totals
30
31  my_dataset = pd.read_table('ETN21_cpx_all.txt')
32  my_dataset = my_dataset[(my_dataset.Total>98) &
33                          (my_dataset.Total<102)]
34  Elements = {'cpx': ['SiO2', 'TiO2', 'Al2O3',
35              'FeO', 'MgO', 'MnO', 'CaO', 'Na2O','Cr2O3']}
36  Cat_Ox_Tolerance = {'cpx': [4,6,0.06]}
37  my_dataset['Tot_cations'] = calc_cations_on_oxygen_basis(
38              myData0 = my_dataset,
39              my_ph = 'cpx',
40              my_el = Elements['cpx'],
41              n_ox = Cat_Ox_Tolerance['cpx'][1])
42
43  my_dataset = my_dataset[(
44      my_dataset['Tot_cations'] < Cat_Ox_Tolerance['cpx'][0] +
45      Cat_Ox_Tolerance['cpx'][2])&(
46      my_dataset['Tot_cations'] > Cat_Ox_Tolerance['cpx'][0] -
47      Cat_Ox_Tolerance['cpx'][2])]
```

Listing 5.1 Initial step of data pre-processing

Moving on in code listing 5.2, we notice that it starts by isolating from the data set the chemical elements in which we are interested (i.e., SiO_2, TiO_2, Al_2O_3, FeO, MgO, CaO, and Na_2O; lines 6–9). The last step of data cleaning consists of removing all rows containing data that are below or exceed the 0.1 and 99.9 percentiles, respectively (lines 11–13).

5.4.2 Compositional Data Analysis (CoDA)

The study of a geochemical data set falls in the field of Compositional Data Analysis (CoDA). In this context, oxides are expressed as a percentage, so their nominal sum is 100%, which defines a "closed" or "compositional" data set (Aitchison, 1982, 1984; Aitchison & Egozcue, 2005). Conducting statistical analysis directly on closed data sets can lead to problems (Aitchison, 1982, 1984; Aitchison & Egozcue, 2005) because some statistical approaches require that the data be normally distributed and not constrained to a constant total value (Boschetty et al., 2022).

```python
from skbio.stats.composition import ilr
from sklearn.preprocessing import RobustScaler
import matplotlib.pyplot as plt
import seaborn as sns

elms_for_clustering  = {'cpx':  ['SiO2', 'TiO2',
           'Al2O3', 'FeO', 'MgO', 'CaO', 'Na2O']}

my_dataset = my_dataset[elms_for_clustering['cpx']]

my_dataset = my_dataset[~((
    my_dataset < my_dataset.quantile(0.001)) |
    (my_dataset > my_dataset.quantile(0.999))).any(axis=1)]

my_dataset_ilr = ilr(my_dataset)

transformer = RobustScaler(
    quantile_range=(25.0, 75.0)).fit(my_dataset_ilr)

my_dataset_ilr_scaled = transformer.transform(my_dataset_ilr)

fig = plt.figure(figsize=(8,8))

for i in range(0,6):
    ax1 = fig.add_subplot(3, 2, i+1)
    sns.kdeplot(my_dataset_ilr_scaled[:, i],fill=True,
         color='k', facecolor='#c7ddf4', ax = ax1)
    ax1.set_xlabel('scaled ilr_' + str(i+1))
fig.align_ylabels()
fig.tight_layout()
```

Listing 5.2 Compositional data analysis (CoDA)

As we know from Sect. 3.3.6, performing multivariate statistical analysis directly on compositional data sets is not formally correct and can bias the results or cause other problems (Aitchison, 1982; Aitchison & Egozcue, 2005; Aitchison, 1984). Different data transformations have been proposed to apply standard and advanced statistical methods to compositional data sets. Examples are the additive log-ratio (*alr*), the centered log-ratio (*clr*), and the isometric log-ratio (*ilr*) transformations (Aitchison, 1982; Aitchison & Egozcue, 2005; Aitchison, 1984). I briefly introduced CoDA analysis in Sect. 3.3.6, where I also presented the equations to perform the *alr*, *clr*, and *ilr* transformations.

At line 15 of code listing 5.2, we apply the *ilr* transformation to our data, then scale in agreement with the median and the inter-quartile range (lines 17–20); that is, we apply *RobustScaler()*. We then visualize the resulting features (Fig. 5.3).

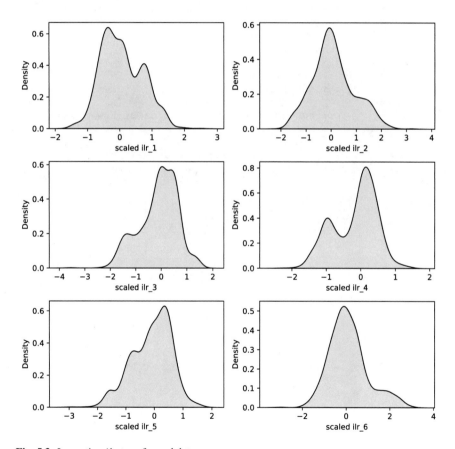

Fig. 5.3 Inspecting *ilr*-transformed data

5.5   Clustering Analyses

Code listing 5.3 shows how to develop a hierarchical clustering dendrogram in
Python (Fig. 5.4). A dendrogram is a tree diagram used to report the result of a
hierarchical clustering estimation (see Sect. 4.4).

```python
1  import numpy as np
2  from sklearn.cluster import AgglomerativeClustering
3  from scipy.cluster.hierarchy import dendrogram,
       set_link_color_palette
4
5  def plot_dendrogram(model, **kwargs):
6
7      counts = np.zeros(model.children_.shape[0])
8      n_samples = len(model.labels_)
9      for i, merge in enumerate(model.children_):
10         current_count = 0
11         for child_idx in merge:
12             if child_idx < n_samples:
13                 current_count +=1
14             else:
15                 current_count += counts[child_idx-n_samples]
16         counts[i] = current_count
17
18     linkage_matrix = np.column_stack([model.children_,
19                                       model.distances_,
20                                       counts]).astype(float)
21
22     dendrogram(linkage_matrix, **kwargs)
23
24 model = AgglomerativeClustering(linkage='ward',
25                                 affinity='euclidean',
26                                 distance_threshold = 0,
27                                 n_clusters=None)
28
29 model.fit(my_dataset_ilr_scaled)
30
31 fig, ax = plt.subplots(figsize = (10,6))
32 ax.set_title('Hierarchical clustering dendrogram')
33
34 plot_dendrogram(model, truncate_mode='level', p=5,
35                 color_threshold=0,
36                 above_threshold_color='black')
37
38 ax.set_xlabel('Number of points in node')
39 ax.set_ylabel('Height')
```

Listing 5.3 Developing a hierarchical clustering dendrogram in Python

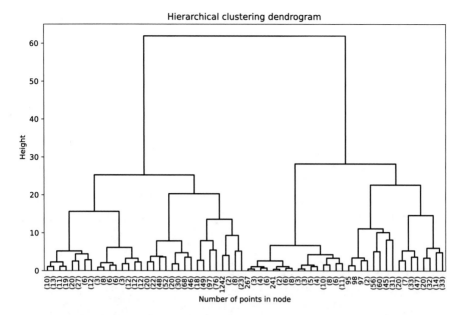

Fig. 5.4 Dendrogram resulting from code listing 5.3

A dendrogram can be oriented either vertically (Fig. 5.4) or horizontally. The orientation can be easily changed in the *dendrogram*() method by using the *orientation* parameter, which takes the values of "top," "bottom," "left," or "right".

```
th = 16.5
fig, ax = plt.subplots(figsize = (10,6))
ax.set_title("Hierarchical clustering dendrogram")
set_link_color_palette(['#000000','#C82127', '#0A3A54',
            '#0F7F8B', '#BFD7EA', '#F15C61', '#E8BFE7'])

plot_dendrogram(model, truncate_mode='level', p=5,
                color_threshold=th,
                above_threshold_color='grey')

plt.axhline(y = th, color = "k", linestyle = "--", lw=1)
ax.set_xlabel("Number of points in node")

fig, ax = plt.subplots(figsize = (10,6))
ax.set_title("Hierarchical clustering dendrogram")
ax.set_ylabel('Height')

plot_dendrogram(model, truncate_mode='lastp', p=6,
                color_threshold=0,
                above_threshold_color='k')

ax.set_xlabel("Number of points in node")
```

Listing 5.4 Refining the dendrogram

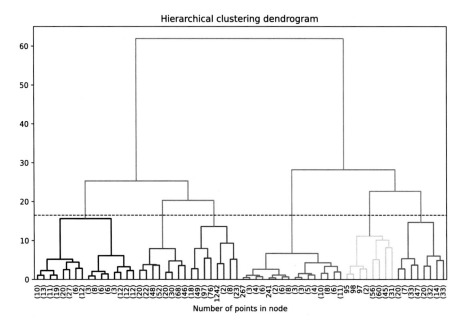

Fig. 5.5 Dendrogram resulting from code listing 5.4

When oriented vertically, the vertical scale gives the distance or similarity between clusters. If we draw a horizontal line, the number of leaves intercepted (see, e.g., Fig. 5.5) defines the number of clusters at that specific height. Increasing the height reduces the number of clusters. In our specific case, fixing a threshold at 16.5 defines six clusters (see code listing 5.4 and Fig. 5.5).

```
1  from sklearn.cluster import AgglomerativeClustering
2  from sklearn.decomposition import PCA
3  import numpy as np
4  import matplotlib.pyplot as plt
5
6  my_colors = {0:'#0A3A54',
7              1:'#E08B48',
8              2:'#BFBFBF',
9              3:'#BD22C6',
10             4:'#FD787B',
11             5:'#67CF62' }
12 #PCA
13 model_PCA = PCA()
14 model_PCA.fit(my_dataset_ilr_scaled)
15 my_PCA = model_PCA.transform(my_dataset_ilr_scaled)
16
17 fig, ax = plt.subplots()
18
```

```
19 ax.scatter(my_PCA[:,0], my_PCA[:,1],
20              alpha=0.6,
21              edgecolors='k')
22
23 ax.set_title('Principal Component Analysys')
24 ax.set_xlabel('PC_1')
25 ax.set_ylabel('PC_2')
```

Listing 5.5 Plotting the first two principal components

5.6 Dimensionality Reduction

The *ilr*-transformed data set consists of six features (Fig. 5.3). To visualize the structure of our data, I performed a Principal Component Analysis (PCA; see Sect. 4.2), which consists of a linear dimensionality reduction that uses a singular value decomposition of the data set to project it onto a lower-dimensional space.

Code listing 5.5 shows how to apply a PCA to our data set. In addition, it provides us with a binary diagram (Fig. 5.6) the shows the two first principal components.

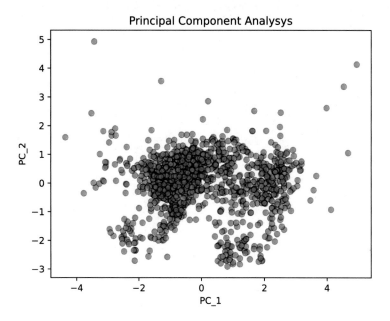

Fig. 5.6 Scatter diagram of the first two principal components

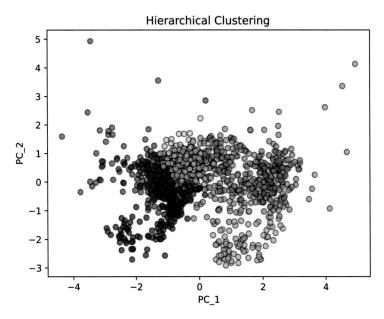

Fig. 5.7 Combining principal component analysis with hierarchical clustering

Visualizing the six clusters highlighted in Fig. 5.5 could be a benefit; code listing 5.6 shows how to do that (Fig. 5.7). Also, code listing 5.6 shows how to apply and visualize (Fig. 5.8) *K*-means clustering (Sect. 4.7).

```
1  #AgglomerativeClustering
2  model_AC = AgglomerativeClustering(linkage='ward',
3                                     affinity='euclidean',
4                                     n_clusters=6)
5  my_AC = model_AC.fit(my_dataset_ilr_scaled)
6
7  fig, ax = plt.subplots()
8  label_to_color = [my_colors[i] for i in my_AC.labels_]
9  ax.scatter(my_PCA[:,0], my_PCA[:,1],
10            c=label_to_color, alpha=0.6,
11            edgecolors='k')
12 ax.set_title('Hierarchical Clustering')
13 ax.set_xlabel('PC_1')
14 ax.set_ylabel('PC_2')
15 my_dataset['cluster_HC'] = my_AC.labels_
16
17 #KMeans
18 from sklearn.cluster import KMeans
19 myKM = KMeans(n_clusters=6).fit(my_dataset_ilr_scaled)
20
21 fig, ax = plt.subplots()
22 label_to_color = [my_colors[i] for i in myKM.labels_]
```

```
23 ax.scatter(my_PCA[:,0], my_PCA[:,1],
24           c=label_to_color, alpha=0.6,
25           edgecolors='k')
26 ax.set_title('KMeans')
27 ax.set_xlabel('PC_1')
28 ax.set_ylabel('PC_2')
29 my_dataset['cluster_KM'] = myKM.labels_
```

Listing 5.6 Combining principal component analysis with hierarchical and *K*-means clustering methods

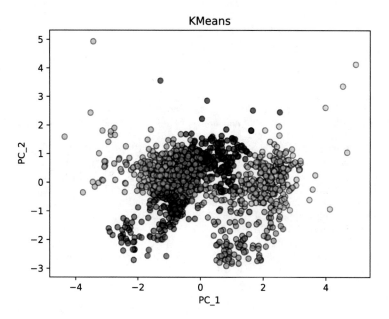

Fig. 5.8 Combining principal component analysis with *K*-means clustering

References

Aitchison, J. (1982). The statistical analysis of compositional data. *Journal of the Royal Statistical Society. Series B (Methodological)*, *44*(2), 139–177.

Aitchison, J. (1984). The statistical analysis of geochemical compositions. *Journal of the International Association for Mathematical Geology*, *16*(6), 531–564.

Aitchison, J., & Egozcue, J. J. (2005). Compositional data analysis: Where are we and where should we be heading? *Mathematical Geology*, *37*(7), 829–850. https://doi.org/10.1007/S11004-005-7383-7

Andronico, D., & Corsaro, R. A. (2011). Lava fountains during the episodic eruption of South-East Crater (Mt. Etna), 2000: Insights into magma-gas dynamics within the shallow volcano plumbing system. *Bulletin of Volcanology, 73*(9), 1165–1178. https://doi.org/10.1007/S00445-011-0467-Y/FIGURES/8

Boschetty, F. O., Ferguson, D. J., Cortés, J. A., Morgado, E., Ebmeier, S. K., Morgan, D. J., Romero, J. E., & Silva Parejas, C. (2022). Insights into magma storage beneath a frequently erupting Arc Volcano (Villarrica, Chile) from unsupervised machine learning analysis of mineral compositions. *Geochemistry, Geophysics, Geosystems, 23*(4), e2022GC010333. https://doi.org/10.1029/2022GC010333

Branca, S., & Del Carlo, P. (2004). Eruptions of Mt. Etna during the past 3,200 years: A revised compilation integrating the historical and stratigraphic records. *Geophysical Monograph Series, 143,* 1–27. https://doi.org/10.1029/143GM02

Cappello, A., Bilotta, G., Neri, M., & Negro, C. D. (2013). Probabilistic modeling of future volcanic eruptions at Mount Etna. *Journal of Geophysical Research: Solid Earth, 118*(5), 1925–1935. https://doi.org/10.1002/JGRB.50190

Caricchi, L., Petrelli, M., Bali, E., Sheldrake, T., Pioli, L., & Simpson, G. (2020b). A data driven approach to investigate the chemical variability of clinopyroxenes from the 2014–2015 Holuhraun–Bárdarbunga eruption (Iceland). *Frontiers in Earth Science,* 8.

Corsaro, R. A., & Miraglia, L. (2022). Near real-time petrologic monitoring on volcanic glass to infer magmatic processes during the February–April 2021 paroxysms of the South-East Crater, Etna. *Frontiers in Earth Science, 10,* 222. https://doi.org/10.3389/FEART.2022.828026/BIBTEX

Costa, F., Shea, T., & Ubide, T. (2020). Diffusion chronometry and the timescales of magmatic processes. *Nature Reviews Earth and Environment, 1*(4), 201–214. https://doi.org/10.1038/s43017-020-0038-x

Di Renzo, V., Corsaro, R. A., Miraglia, L., Pompilio, M., & Civetta, L. (2019). Long and short-term magma differentiation at Mt. Etna as revealed by Sr-Nd isotopes and geochemical data. *Earth-Science Reviews, 190,* 112–130. https://doi.org/10.1016/J.EARSCIREV.2018.12.008

Ferlito, C., Coltorti, M., Lanzafame, G., & Giacomoni, P. P. (2014). The volatile flushing triggers eruptions at open conduit volcanoes: Evidence from Mount Etna volcano (Italy). *Lithos, 184–187,* 447–455. https://doi.org/10.1016/J.LITHOS.2013.10.030

Higgins, O., Sheldrake, T., & Caricchi, L. (2021). Machine learning thermobarometry and chemometry using amphibole and clinopyroxene: a window into the roots of an arc volcano (Mount Liamuiga, Saint Kitts). *Contributions to Mineralogy and Petrology, 177*(1), 1–22. https://doi.org/10.1007/S00410-021-01874-6

Jorgenson, C., Higgins, O., Petrelli, M., Bégué, F., & Caricchi, L. (2022). A machine learning-based approach to clinopyroxene thermobarometry: Model optimization and distribution for use in Earth Sciences. *Journal of Geophysical Research: Solid Earth, 127*(4), e2021JB022904. https://doi.org/10.1029/2021JB022904

Musu, A., Corsaro, R. A., Higgins, O., Jorgenson, C., Petrelli, M., & Caricchi, L. (2023). The magmatic evolution of South-East Crater (Mt. Etna) during the February-April 2021 sequence of lava fountains from a mineral chemistry perspective. *Bulletin of Volcanology, 85,* 33.

Okrusch, M., & Frimmel, H. E. (2020). *Mineralogy.* Springer Berlin Heidelberg. https://doi.org/10.1007/978-3-662-57316-7

Paffengoltz, K. N. (1978). *Geological dictionary.* Nedra Publishing.

Petrelli, M., Caricchi, L., & Perugini, D. (2020). Machine learning thermo-barometry: Application to clinopyroxene-bearing magmas. *Journal of Geophysical Research: Solid Earth, 125*(9). https://doi.org/10.1029/2020JB020130

Petrelli, M., & Zellmer, G. (2020). *Rates and timescales of magma transfer, storage, emplacement, and eruption.* https://doi.org/10.1002/9781119521143.ch1

Putirka, K. (2008). Thermometers and barometers for volcanic systems. https://doi.org/10.2138/rmg.2008.69.3

Ubide, T., & Kamber, B. (2018). Volcanic crystals as time capsules of eruption history. *Nature Communications, 9*(1). https://doi.org/10.1038/s41467-017-02274-w

Ubide, T., Neave, D., Petrelli, M., & Longpré, M.-A. (2021). Editorial: Crystal archives of magmatic processes. *Frontiers in Earth Science, 9.* https://doi.org/10.3389/feart.2021.749100

Chapter 6
Clustering of Multi-Spectral Data

6.1 Spectral Data from Earth-Observing Satellites

Earth-observing satellite missions such as Sentinel[1] and Landsat[2] provide us with multispectral, hyperspectral, and panchromatic data. The Sentinel earth-observing satellite missions are part of the Copernicus program, developed by the European Space Agency,[3] whereas the Landsat Program is jointly managed by NASA and the U.S. Geological Survey (see footnote 2).

Spectral images are two-dimensional representations of surface reflectance or radiation in different bands of the electromagnetic spectrum. Multi-spectral and hyper-spectral data are acquired by multiple sensors operating over wide and narrow (sometimes quasi-continuous) wavelength ranges, respectively. In contrast, panchromatic images are acquired by detectors covering the entire visible range.

Multi-spectral, hyper-spectral, and panchromatic data can be combined and modulated to produce new indexes[4] (e.g., the Generalized Difference Vegetation Index or the Normalized Difference Snow Index), which highlight specific phenomena and facilitate data interpretation.

For example, the Sentinal-2 Multi-spectral Instrument operates over 13 spectral bands. Four bands labeled B2, B3, B4, and B8 provide a spatial resolution of 10 m, six bands labeled B5, B6, B7, B8a, B11, and B12 provide a spatial resolution of

---

[1] https://sentinels.copernicus.eu.

[2] https://landsat.gsfc.nasa.gov.

[3] https://www.esa.int.

[4] https://www.usgs.gov/landsat-missions/landsat-surface-reflectance-derived-spectral-indices.

© The Author(s), under exclusive license to Springer Nature Switzerland AG 2023
M. Petrelli, *Machine Learning for Earth Sciences*, Springer Textbooks
in Earth Sciences, Geography and Environment,
https://doi.org/10.1007/978-3-031-35114-3_6

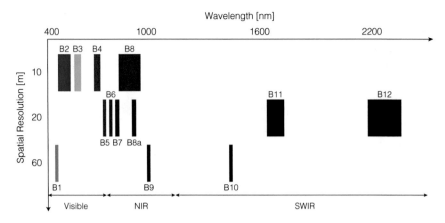

Fig. 6.1 Spectral bands of Sentinel2 satellites. Modified from Majidi Nezhad et al. (2021)

20 m, and three bands labeled B1, B9, and B10 provide a spatial resolution of 60 m (Fig. 6.1).

6.2   Import Multi-Spectral Data to Python

Multi-spectral data can be downloaded from numerous access points, such as the USGS Earth Explorer,[5] the Copernicus Open Access Hub,[6] and Theia.[7]

As an example, consider Fig. 6.2, which represents the recombination of the B4, B3, and B2 bands to form a RGB (i.e., red, green, blue) image from a Sentinel2 acquisition downloaded from the Theia portal. The image location is southern New South Wales (Australia).[8] Each side of the square image measures about 110 km.

Figure 6.3 shows the data structure of a Sentinel2 repository downloaded from Theia. The repository follows the MUSCATE[9] nomenclature and contains a metadata file, a quick-look file, numerous Geo-Tiff image files, and two sub-repositories MASKS and DATA, which contain supplementary data. The naming enables us to uniquely identify each product and consists of many tags, starting with a platform identification (i.e., Sentinel2B) followed by the date of acquisition in the format YYYYMMDD-HHmmSS-sss (e.g., 20210621-001635-722), with YYYY being the year, MM the month, DD the day, HH the hour in 24 hour format, mm the minutes, SS the seconds, and sss the milliseconds. The subsequent tags refer to product level (e.g., L2A), geographical zone (e.g., T55HDB_C), and product version (e.g., V2-2). The letter L, a number, and another letter characterize different product levels with the exception of level L0, which is compressed raw data and is not followed by any letter. Levels L1A, L1B, and L2A correspond to uncompressed

---

[5] https://earthexplorer.usgs.gov.

[6] https://scihub.copernicus.eu.

[7] https://catalogue.theia-land.fr.

[8] https://bit.ly/ml_geart.

[9] https://www.theia-land.fr/en/product/sentinel-2-surface-reflectance/.

Fig. 6.2 RGB composite image where the B4, B3, and B2 bands regulate the intensities of the red, green, and blue channels, respectively

raw data, radiometrically corrected radiance data, and orthorectified bottom-of-atmosphere reflectance, respectively.[10] Spectral Geo-Tiff files also use an additional tag, namely, SRE and FRE, which correspond respectively to images taken in ground reflectance without correcting for slope effects and images taken in ground reflectance with slope effects corrected. We shall work on FRE data.

To import Sentinel2 multi-spectral data, I used Rasterio,[11] which is a Python API based on Numpy and GeoJSON (i.e., an open standard format designed for representing geographical features, along with their non-spatial attributes) to read, write, and manage Geo-Tiff data.

---

[10] https://sentinels.copernicus.eu/web/sentinel/technical-guides/sentinel-2-msi.

[11] https://rasterio.readthedocs.io/.

Fig. 6.3 Sentinel2 data structure

If you followed the instructions in Chap. 2, your Python machine learning environments named *env_ml* and *env_ml_intel* already contain Rasterio. With Rasterio, opening Geo-Tiff files is straightforward (see code listing 6.1).

```
 1  import rasterio
 2  import numpy as np
 3
 4  imagePath = 'SENTINEL2B_20210621-001635-722_L2A_T55HDB_C_V2-2/
        SENTINEL2B_20210621-001635-722_L2A_T55HDB_C_V2-2_FRE_'
 5
 6  bands_to_be_inported = ['B2', 'B3', 'B4', 'B8']
 7
 8  bands_dict = {}
 9  for band in bands_to_be_inported:
10      with rasterio.open(imagePath+ band +'.tif', 'r',
11                         driver='GTiff') as my_band:
12          bands_dict[band] = my_band.read(1)
```

Listing 6.1 Using Rasterio to import Sentinel2 data in Python

Code listing 6.1 creates a dictionary of NumPy arrays (i.e., *bands_dict*) containing spectral information from B2, B3, B4, and B8 corresponding to the blue, green, red, and near-infrared bands, respectively. In code listing 6.1, we limit the import to four bands, all acquired at the same spatial resolution of 10 m. However, the script can be easily extended to import more bands.

Combining the data from the *bands_dict* dictionary allows many different representations to be achieved. For example, Sovdat et al. (2019) explain how to obtain the "natural color" representation of Sentinel-2 data.

Explaining how to obtain a perfectly balanced image with natural colors is beyond the scope of this book, so we limit ourselves to combining bands B2, B3, and B4, which roughly correspond to blue, green, and red as perceived by our eyes.

In detail, a bright, possibly overly saturated (Sovdat et al., 2019) image (i.e., *r_g_b*) can be easily derived and plotted (see code listing 6.2 and Fig. 6.2). We start from the *bands_dict* dictionary after contrast stretching (lines 11–17) and scale the values in the interval [0,1]. This is the so-called "true color" representation. Sometimes, bands B3 (red) and B4 (green) are combined with B8 (near-infrared) to achieve a "false color'" representation. False color composite images are often used to highlight plant density and health (see, e.g., Fig. 6.4). Code listing 6.3 shows how to construct a false-color representation (i.e., *nir_r_g)* of Sentinel2 data.

```
1  import numpy as np
2  from skimage import exposure, io
3  from skimage.transform import resize
4  import matplotlib.pyplot as plt
5
6  r_g_b = np.dstack([bands_dict['B4'],
7                     bands_dict['B3'],
8                     bands_dict['B2']])
9
10 # contrast stretching and rescaling between [0,1]
11 p2, p98 = np.percentile(r_g_b, (2,98))
12 r_g_b = exposure.rescale_intensity(r_g_b, in_range=(p2, p98))
13 r_g_b = r_g_b / r_g_b.max()
14
15 fig, ax = plt.subplots(figsize=(8, 8))
16 ax.imshow(r_g_b)
17 ax.axis('off')
```

Listing 6.2 Plotting a RGB image using bands B4, B3, and B2

```
1  import numpy as np
2  from skimage import exposure, io
3  from skimage.transform import resize
4  import matplotlib.pyplot as plt
5
6  nir_r_g = np.dstack([bands_dict['B8'],
7                       bands_dict['B4'],
8                       bands_dict['B3']])
9
10 # contrast stretching and rescaling between [0,1]
11 p2, p98 = np.percentile(nir_r_g, (2,98))
```

```
12 nir_r_g = exposure.rescale_intensity(nir_r_g, in_range=(p2, p98))
13
14 fig, ax = plt.subplots(figsize=(8, 8))
15 ax.imshow(nir_r_g)
16 ax.axis('off')
```

Listing 6.3 Plotting a false-color RGB composite image using bands B8, B4, and B3

6.3 Descriptive Statistics

One of the first steps of any ML workflow deals with descriptive statistics. For our
Sentinel2 data set, code listing 6.4 shows how to obtain descriptive statistics via
the visualization of a four-band (i.e., B2, B3, B4, and B5) array derived from Geo-

Fig. 6.4 Image resulting from code listing 6.3

Tiff data. On line 5, we create a (10 980, 10 980, 4) array (i.e., the *my_array_2d* characterized by a width, height, and depth of 10 980, 10 980, and 4, respectively) from the dictionary created in code listing 6.1. In the next step (line 10), we create a new array (*my_array_1d*) that reshapes *my_array_2d* from (10 980, 10 980, 4) to (120 560 400, 4). This is the typical dimensions of an array that is ready for ML processing in scikit-learn. Converting *my_array_1d* to a pandas DataFrame (i.e., *my_array_1d_pandas*) facilitates the visualization (see lines 18–46) and produces the most basic descriptive statistics (i.e., listing 6.5). Code listing 6.5 reveals basic information about the central tendency, dispersion, and shape of our input features.

```python
 1 import numpy as np
 2 import matplotlib.pyplot as plt
 3 import pandas as pd
 4
 5 my_array_2d = np.dstack([bands_dict['B2'],
 6                          bands_dict['B3'],
 7                          bands_dict['B4'],
 8                          bands_dict['B8']])
 9
10 my_array_1d =my_array_2d[:,:,:4].reshape(
11     (my_array_2d.shape[0] * my_array_2d.shape[1],
12      my_array_2d.shape[2]))
13
14 my_array_1d_pandas = pd.DataFrame(my_array_1d,
15                     columns=['B2', 'B3', 'B4', 'B8'])
16
17
18 fig, (ax1, ax2) = plt.subplots(1, 2, figsize=(7,3))
19 my_medianprops = dict(color='#C82127', linewidth = 1)
20 my_boxprops = dict(facecolor='#BFD7EA', edgecolor='#000000')
21 ax1.boxplot(my_array_1d_pandas, vert=False, whis=(0.5, 99.5),
22             showfliers=False, labels=my_array_1d_pandas.columns,
23             patch_artist=True, showcaps=False,
24             medianprops=my_medianprops, boxprops=my_boxprops)
25 ax1.set_xlim(-0.1,0.5)
26 ax1.set_xlabel('Surface reflectance Value')
27 ax1.set_ylabel('Band Name')
28 ax1.grid()
29 ax1.set_facecolor((0.94, 0.94, 0.94))
30
31 colors=['#BFD7EA','#0F7F8B','#C82127','#F15C61']
32 for band, color in zip(my_array_1d_pandas.columns, colors):
33     ax2.hist(my_array_1d_pandas[band], density=True,
34              bins='doane', range=(0,0.5), histtype='step',
35              linewidth=1, fill=True, color=color, alpha=0.6,
36              label=band)
37     ax2.hist(my_array_1d_pandas[band], density=True,
38              bins='doane', range=(0,0.5), histtype='step',
39              linewidth=0.5, fill=False, color='k')
40 ax2.legend(title='Band Name')
41 ax2.set_xlabel('Surface Reflectance Value')
```

```
42 ax2.set_ylabel('Probability Density')
43 ax2.xaxis.grid()
44 ax2.set_facecolor((0.94, 0.94, 0.94))
45 plt.tight_layout()
46 plt.savefig('descr_stat_sat.pdf')
```

Listing 6.4 Descriptive statistics and data visualization

For example, Fig. 6.5 shows that 99% of the reflectance data for B2, B3, B4, and B8 fall in the range 0.015–0.420. However, maximum values are always greater than unity (i.e., the upper theoretical bound for reflectance data). Outliers with reflectance values greater than unity could be the result of specular effects due to surfaces or clouds (Schaepman-Strub et al., 2006).

```
In [1]: my_array_1d_pandas.describe().applymap("{0:.3f}".format)
Out[1]:
```

	B2	B3	B4	B8
count	120560400.000	120560400.000	120560400.000	120560400.000
mean	0.042	0.062	0.076	0.186
std	0.013	0.016	0.026	0.056
min	0.000	0.000	0.000	0.000
25%	0.035	0.053	0.061	0.151
50%	0.042	0.062	0.076	0.177
75%	0.049	0.070	0.091	0.210
max	1.443	1.304	1.277	1.201

Listing 6.5 Descriptive statistics using pandas *describe()*

If not addressed correctly, large outliers could affect the results of your ML model. Consequently, I suggest implementing a strategy to remove the outliers based on robust statistics (see, e.g., Petrelli, 2021) or applying a robust scaler (cf. paragraph 3.3.5).

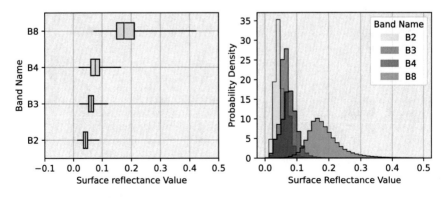

Fig. 6.5 Descriptive statistics resulting from code listing 6.4

6.4 Pre-processing and Clustering

This section presents a simplified workflow to cluster our Sentinel2 data. As input features, I used *my_array_1d*, which contains reflectance data from B2, B3, B4, and B8. Note that many different strategies are reported in the literature for selecting input features, such as using band ratios, specific indexes, or combinations of bands, band ratios, and indexes (e.g., Ge et al., 2020). Due to the presence of large outliers, I opted for the *RobustScaler*() algorithm (line 6 of code listings 6.6 and 6.7) in scikit-learn.

```python
1 from sklearn.preprocessing import  RobustScaler
2 from sklearn import cluster
3 import matplotlib.colors as mc
4 import matplotlib.pyplot as plt
5
6 X = RobustScaler().fit_transform(my_array_1d)
7 my_ml_model = cluster.KMeans(n_clusters=5)
8 learning = my_ml_model.fit(X)
9 labels_1d = learning.labels_
10
11 labels_1d = my_ml_model.predict(X)
12 labels_2d = labels_1d.reshape(my_array_2d[:,:,0].shape)
13
14 cmap = mc.LinearSegmentedColormap.from_list("", ["black","red","
      yellow", "green", "blue"])
15 fig, ax = plt.subplots(figsize=[18,18])
16 ax.imshow(labels_2d, cmap=cmap)
17 ax.axis('off')
```

Listing 6.6 Implementing *K*-means clustering

```python
1 from sklearn.preprocessing import  RobustScaler
2 from sklearn import  mixture
3 import matplotlib.colors as mc
4 import matplotlib.pyplot as plt
5
6 X = RobustScaler().fit_transform(my_array_1d)
7 my_ml_model = mixture.GaussianMixture(n_components=5,
      covariance_type="full")
8 labels_1d = my_ml_model.predict(X)
9
10 labels_2d = labels_1d.reshape(my_array_2d[:,:,0].shape)
11
12 cmap = mc.LinearSegmentedColormap.from_list("", ["black","red","
      yellow", "green","blue"])
13 fig, ax = plt.subplots(figsize=[18,18])
14 ax.imshow(labels_2d, cmap=cmap)
15 ax.axis('off')
```

Listing 6.7 Implementing Gaussian mixture clustering

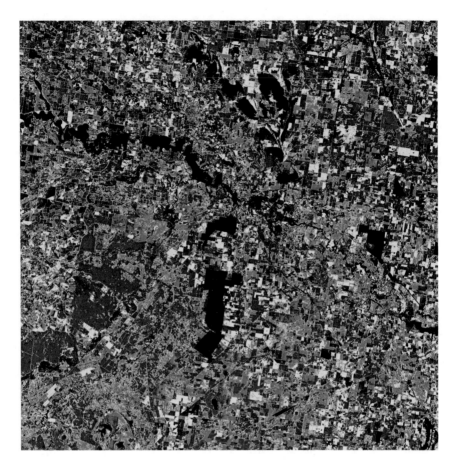

Fig. 6.6 K-means clustering. Image resulting from code listing 6.6

For the first attempt at clustering (code listing 6.6), I selected the K-means algorithm, fixing the number of clusters to five (line 7). I then started the unsupervised
learning at line 8. Lines 11 and 12 collect the labels (i.e., a number from 0 to 4)
assigned by the K-means algorithm to each element (i.e., to each pixel of the image)
of *my_array_1d* and I reported the elements in the same two-dimensional geometry
of the original image (Fig. 6.2). Finally, the different clusters using different colors
(i.e. lines 14–17) are plotted in Fig. 6.6.

For the second attempt at clustering (code listing 6.7), I selected the Gaussian
mixtures algorithm, again fixing the number of clusters to five (line 7). Figure 6.7
shows the clustering result obtained by the Gaussian mixture algorithm.

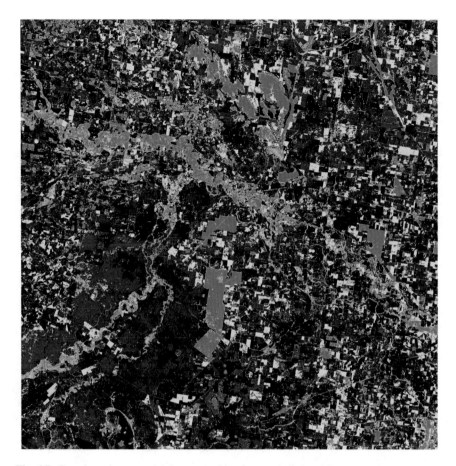

Fig. 6.7 Gaussian mixture model. Image resulting from code listing 6.7

References

Ge, W., Cheng, Q., Jing, L., Wang, F., Zhao, M., & Ding, H. (2020). Assessment of the capability of sentinel-2 imagery for iron-bearing minerals mapping: A case study in the cuprite area, Nevada. *Remote Sensing, 12*(18), 3028. https://doi.org/10.3390/RS12183028

Majidi Nezhad, M., Heydari, A., Pirshayan, E., Groppi, D., & Astiaso Garcia, D. (2021). A novel forecasting model for wind speed assessment using sentinel family satellites images and machine learning method. *Renewable Energy, 179*, 2198–2211. https://doi.org/10.1016/J.RENENE.2021.08.013

Petrelli, M. (2021). *Introduction to python in earth science data analysis*. Berlin: Springer. https://doi.org/10.1007/978-3-030-78055-5

Schaepman-Strub, G., Schaepman, M. E., Painter, T. H., Dangel, S., & Martonchik, J. V. (2006). Reflectance quantities in optical remote sensing—definitions and case studies. *Remote Sensing of Environment, 103*(1), 27–42. https://doi.org/10.1016/J.RSE.2006.03.002

Sovdat, B., Kadunc, M., Batič, M., & Milčinski, G. (2019). Natural color representation of Sentinel-2 data. *Remote Sensing of Environment, 225*, 392–402. https://doi.org/10.1016/J.RSE.2019.01.036

Part III
Supervised Learning

Chapter 7
Supervised Machine Learning Methods

7.1 Supervised Algorithms

To learn, supervised algorithms use the labels (i.e., the solutions) that appear in the training data set. This chapter introduces the supervised ML algorithms for regression and classification that are shown in Fig. 3.5. In addition, specific references are given for those who wish to go deeper into the mathematics behind these ML methods.

7.2 Naive Bayes

Since Bayesian statistics is rarely introduced to geology students, I introduce Bayes theorem here before describing how it is applied in ML (e.g., naive Bayes).

Probabilities Figure 7.1 describes a simplified set of rock textures containing $n_{tot} = 10$ elements. The set comes from six porphyritic, one holocrystalline, and three aphyric igneous rocks. The probability $P(ol)$ of randomly picking a rock containing olivines is thus 3/10. In Bayesian statistical inference, the probability $P(ol)$ is called the "prior probability," which is the probability of an event before new data are collected.

Conditional Probabilities Assume now that we want to know the probability of picking a rock containing olivines if we pick a rock characterized by a dark matrix. In this case, the conditional probability $P(ol|dark) = 1/3$.

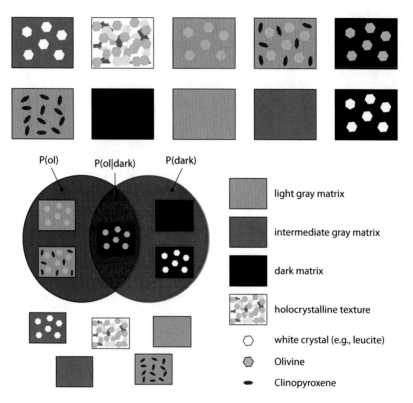

Fig. 7.1 Understanding conditional probabilities and Bayes formulation

Joint Probabilities Please keep in mind that the term "conditional probability" is not a synonym of "joint probability" and these two concepts should not be confused. Also, be sure to use the correct notation: for joint probability, the terms are separated by commas [e.g., $P(\text{ol, dark})$], whereas for conditional probability, the terms are separated by a vertical bar [e.g., $P(\text{ol}|\text{dark})$]. Note that $P(\text{ol, dark})$ is the probability of randomly picking a rock that contains olivines and is characterized by a dark matrix [i.e., $P(\text{ol, dark}) = 1/10$]. In contrast, $P(\text{ol}|\text{dark})$ is the probability of a rock containing olivines from among those that have a dark matrix, $P(\text{ol}|\text{dark}) = 1/3$. Joint probabilities and conditional probabilities are related as follows:

$$P(\text{ol, dark}) = P(\text{ol}|\text{dark})P(\text{dark}). \qquad (7.1)$$

Deriving the Bayes Formulation As in Eq. (7.1), we could write

$$P(\text{dark, ol}) = P(\text{dark}|\text{ol})P(\text{ol}). \qquad (7.2)$$

Since $P(\text{dark, ol}) = P(\text{ol, dark})$, the right-hand terms of Eqs. (7.1) and (7.2) must be equal:

$$P(\text{dark|ol})P(\text{ol}) = P(\text{ol|dark})P(\text{dark}). \tag{7.3}$$

Dividing both sides of Eq. (7.3) by $P(\text{ol})$, we get Bayes formula for our specific case:

$$P(\text{dark|ol}) = \frac{P(\text{ol|dark})P(\text{dark})}{P(\text{ol})}. \tag{7.4}$$

Generalizing Eq. (7.4), we get the well-known Bayes equation:

$$P(A|B) = \frac{P(B|A)P(A)}{P(B)}. \tag{7.5}$$

Naive Bayes for Classification To understand the naive Bayes ML algorithm, I propose the same workflow as described by Zhang (2004). Assume that you want to classify a set $X = (x_1, x_2, x_3, \ldots, x_n)$ and that c is the label of your class. For simplicity, assume that c is strictly positive (+) or negative (−); in other words, we have only two classes. In this case, the Bayes formula takes the form

$$P(c|X) = \frac{P(X|c)P(c)}{P(X)}. \tag{7.6}$$

X is classified as being in class $c = +$ if and only if

$$f_b(X) = \frac{P(c = +|X)}{P(c = -|X)} \geq 1, \tag{7.7}$$

where $f_b(X)$ is the Bayesian classifier.

Now assume that all the features are independent (i.e., the naive assumption). We can write

$$P(X|c) = P(x_1, x_2, x_3, \ldots, x_n|c) = \prod_{i=1}^{n} P(x_i|c). \tag{7.8}$$

The resulting naive Bayesian classifier $f_{nb}(X)$, or simply "naive Bayes" classifier, can be written as

$$f_{nb}(X) = \frac{P(c = +)}{P(c = -)} \prod_{i=1}^{n} \frac{P(x_i|c = +)}{P(x_i|c = -)}. \tag{7.9}$$

Note that the naive assumption is a strong constraint that, in Earth Sciences, is often violated. If feature independence is violated, we have two options: The first is to estimate $P(X|c)$ without using the naive assumption (Kubat, 2017). However, using this option inevitably increases the complexity of the problem (Kubat, 2017). The second option is more pragmatic: we reduce the feature dependence by appropriate data pre-processing. As suggested by Kubat (2017), a starting point is to avoid using redundant features.

In scikit-learn the *GaussianNB()* method implements the Gaussian naive Bayes algorithm for classification with $P(X|c)$ assumed to be multivariate normal distributed.

7.3 Quadratic and Linear Discriminant Analysis

Like naive Bayes, quadratic and linear discriminant analyses (QDA and LDA, respectively) rely on the Bayes theorem. Assume that $f_c(x)$ is the class-conditional density of X in class c, and let π_c be the prior probability of class c, with $\sum_{c=1}^{K} \pi_c = 1$, where K is the number of classes. The Bayes theorem states (Kubat, 2017)

$$P(c|X) = \frac{f_c(x)\pi_c}{\sum_{l=1}^{K} f_l(x)\pi_l}. \tag{7.10}$$

Now, modeling each class density as multivariate Gaussian,

$$f_c(x) = \frac{1}{(2\pi)^{p/2} \left|\sum_c\right|^{1/2} e^{-\frac{1}{2}(x-\mu_c)^T \sum_c^{-1}(x-\mu_c)}}, \tag{7.11}$$

we define the QDA. The LDA constitutes a special case of the QDA if the classes have a common covariance matrix (i.e., $\sum_c = \sum \forall c$). The main difference between LDA and QDA depends on the resulting decision boundaries being linear or quadratic functions, respectively.

The algorithms for LDA and QDA are similar, except that separate covariance matrices must be estimated for each class in QDA. Given a large number of features, this implies a dramatic increase in the computed parameters. For K classes and p features, LDA and QDA compute $(K-1)x(p+1)$ and $(K-1)x[p(p+3)/2+1]$ parameters, respectively. In scikit-learn, the methods *LinearDiscriminantAnalysis()* and *QuadraticDiscriminantAnalysis()* perform LDA and QDA, respectively.

7.4 Linear and Nonlinear Models

Sugiyama (2015) defines d-dimensional linear-in-parameter models as

$$f_\theta(x) = \sum_{j=i}^{b} \theta_j \phi_j(x) = \theta^T \phi(x), \tag{7.12}$$

where x, ϕ, and θ are a d-dimensional input vector, a basis function, and the parameters of the basis function, respectively, and b is the number of basis functions. As an example, given a one-dimensional input, Eq.(7.12) reduces to (Sugiyama, 2015)

$$f_\theta(x) = \sum_{j=i}^{b} \theta_j \phi_j(x) = \theta^T \phi(x), \tag{7.13}$$

where

$$\phi(x) = (\phi_1(x), \ldots, \phi_b(x))^T, \tag{7.14}$$

and

$$\theta = (\theta_1, \ldots, \theta_b)^T. \tag{7.15}$$

Note that linear-in-parameter models are linear in θ and can handle straight lines (i.e., linear-in-input models such as code listing 7.1 and Fig. 7.2):

$$\phi(x) = (1, x)^T, \tag{7.16}$$

$$\theta = (\theta_1, \theta_2)^T. \tag{7.17}$$

Linear-in-parameter models can also manage nonlinear functions such as polynomials (e.g., code listing 7.1 and Fig. 7.2):

$$\phi(x) = (1, x, x^2, \ldots, x^{b-1})^T, \tag{7.18}$$

$$\theta = (\theta_1, \theta_2, \ldots, \theta_b)^T. \tag{7.19}$$

```python
import numpy as np
import matplotlib.pyplot as plt

x = np.arange(1,6)
y = np.array([0,1,2,9,9])

fig, ax = plt.subplots()
ax.scatter(x, y, marker = 'o', s = 100, color = '#c7ddf4',
    edgecolor = 'k')

orders = np.array([1,2,4])
colors =['#ff464a','#342a77','#4881e9']
linestiles = ['-','--','-.']

for order, color, linestile in zip(orders, colors, linestiles):
```

```
15    betas = np.polyfit(x, y, order)
16    func = np.polyld(betas)
17    x1 = np.linspace(0.5,5.5, 1000)
18    y1 = func(x1)
19    ax.plot(x1, y1, color=color, linestyle=linestile, label="
      Linear-in-parameters model of order " + str(order))
20
21 ax.legend()
22 ax.set_xlabel('A quantity relevant in geology\n(e.g., time)')
23 ax.set_ylabel('A quantity relevant in geology\n(e.g., spring flow
      rate)')
24 fig.tight_layout()
```

Listing 7.1 Polynomial regression as example of linear-in-parameter modeling

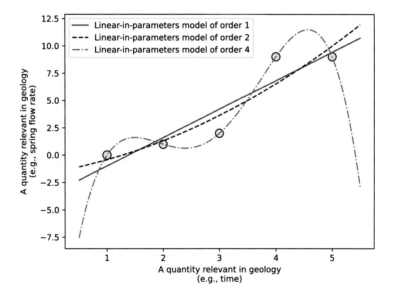

Fig. 7.2 Result of code listing 7.1

Given an input vector x of p values, linear-in-parameter models can still manage linear-in-input problems, such as managing hyper-planes:

$$\phi(x) = (1, x_1, x_2, \ldots, x_p)^T, \tag{7.20}$$

$$\theta = (\theta_1, \theta_2, \ldots, \theta_b)^T. \tag{7.21}$$

In this case, the number of basis functions corresponds to the dimension of the input vector plus one (i.e., $b = p+1$). Some authors prefer to report the first term of

$\boldsymbol{\theta}$ separately, calling it the "bias" (i.e., θ_0), and reformulating the problem as follows:

$$\boldsymbol{\phi}(\boldsymbol{x}) = (x_1, x_2, \ldots, x_{b=p})^T, \tag{7.22}$$

$$\boldsymbol{\theta} = (\beta_0, \boldsymbol{\beta}), \tag{7.23}$$

with

$$\boldsymbol{\beta} = (\beta_1, \beta_2, \ldots, \beta_{b=p},)^T. \tag{7.24}$$

All $f_{\boldsymbol{\theta}}(\boldsymbol{x})$ models that cannot be expressed as linear in their parameters fall in the field of nonlinear modeling (Sugiyama, 2015).

7.5   Loss Functions, Cost Functions, and Gradient Descent

Most ML algorithms involve model optimization [e.g., $f_{\boldsymbol{\theta}}(\boldsymbol{x})$ in Eq. (7.13)]. For the purposes of this book, the term "optimization" shall refer to adjusting the model parameters $\boldsymbol{\theta}$ to minimize or maximize a function that measures the consistency between model predictions and training data.

In general, the function we want to minimize or maximize is called the **objective function** (Goodfellow et al., 2016). In the case of minimization, the objective function takes names such as cost function, loss function, and error function. These terms are often interchangeable Goodfellow et al. (2016), but sometimes a specific term is used such as loss or cost function to describe a specific task.

As an example, some authors use the term **loss function** to measure how well a model agrees with a single label in the training data set (Goodfellow et al., 2016). The square loss is an example of a loss function:

$$L(\boldsymbol{\theta}) = [y_i - f_{\boldsymbol{\theta}}(\mathbf{x}_i)]^2, \tag{7.25}$$

where y_i and $f_{\boldsymbol{\theta}}(\mathbf{x}_i)$ are the labeled (i.e., true or measured) values and those predicted by our model, respectively. Also, $\mathbf{x}$ and $\boldsymbol{\theta}$ are the inputs and the parameters governing the model, respectively.

Similarly, the **cost function** evaluates the loss function over the entire data set and helps to evaluate the overall performance of the model (Goodfellow et al., 2016). The mean squared error is an example of a cost function:

$$C(\boldsymbol{\theta}) = \frac{1}{n} \sum_{i=1}^{n} [y_i - f_{\boldsymbol{\theta}}(\mathbf{x}_i)]^2, \tag{7.26}$$

where n is the number of elements in the training data set.

Typically, our aim is to minimize the cost function $C(\boldsymbol{\theta})$, and the gradient descent (GD) is an appropriate method to do this. GD works by updating the parameters (in our case $\boldsymbol{\theta}$) governing our model [i.e., $f_{\boldsymbol{\theta}}(\mathbf{x})$], in the direction opposite that of the cost-function gradient $\nabla C(\boldsymbol{\theta})$ (Sugiyama, 2015):

$$\boldsymbol{\theta}^{t+1} = \boldsymbol{\theta}^t - \gamma \nabla C(\boldsymbol{\theta}^t). \tag{7.27}$$

In the simplest example of linear regression with $\mathbf{x} \in \mathbb{R}$,

$$f_{\boldsymbol{\theta}}(\mathbf{x}) = \theta_1 + \theta_2 x, \tag{7.28}$$

the mean squared-error cost function is

$$C(\boldsymbol{\theta}) = \frac{1}{n} \sum_{i=1}^{n} [y_i - (\theta_1 + \theta_2 x_i)]^2. \tag{7.29}$$

Note that the simple linear example in $\mathbb{R}$ can be easily generalized to $\mathbb{R}^d$. Also, note that the example of linear regression proposed here has a well-known and easy-to-apply least squares analytical solution in the case of linearity (i.e., x is linear in the mean of y), independence (i.e., the observations are independent of each other), and normality [i.e., for any fixed value of x, y is normally distributed (Maronna et al., 2006)]. However, this self-explanatory example shows how GD works.

To develop a GD, the first step is to compute the partial derivative of $C(\boldsymbol{\theta})$ with respect to θ_1 and θ_2. Therefore, we write

$$D_{\theta_1} = \frac{-2}{n} \sum_{i=1}^{n} [y_i - (\theta_1 + \theta_2 x_i)], \tag{7.30}$$

$$D_{\theta_2} = \frac{-2}{n} \sum_{i=1}^{n} [y_i - (\theta_1 + \theta_2 x_i)] x_i. \tag{7.31}$$

```
1  import numpy as np
2  import matplotlib.pyplot as plt
3  line_colors = ['#F15C61','#0F7F8B','#0A3A54','#C82127']
4
5  # linear data set with noise
6  n = 100
7  theta_1, theta_2 = 3, 1 # target value for theta_1 & theta_2
8  x = np.linspace(-10, 10, n)
9  np.random.seed(40)
10 noise = np.random.normal(loc=0.0, scale=1.0, size=n)
11 y = theta_1 + theta_2 * x + noise
12 fig, (ax1, ax2) = plt.subplots(2, 1, figsize=(6, 12))
13 ax1.scatter(x, y, c='#BFD7EA',  edgecolor='k')
14
```

```
15 my_theta_1, my_theta_2  = 0, 0 # arbitrary initial values
16 gamma = 0.0005 # learning rate
17 t_final = 10001 # umber of itrations
18 n = len(x)
19 to_plot,  cost_function = [1, 25, 500, 10000], []
20 # Gradient Descent
21 for i in range(t_final):
22       #Eq. 4.30
23       D_theta_1 = (-2/n)*np.sum(y-(my_theta_1 + my_theta_2*x))
24       #Eq. 4.31
25       D_theta_2 = (-2/n)*np.sum(x*(y-(my_theta_1+my_theta_2*x)))
26
27       my_theta_1 = my_theta_1 - gamma * D_theta_1 #Eq. 4.32
28       my_theta_2 = my_theta_2 - gamma * D_theta_2 #Eq. 4.33
29       cost_function.append(((1/n)  * np.sum(y - (my_theta_1 +
         my_theta_2 * x))**2)
30
31       if i in to_plot:
32           color_index = to_plot.index(i)
33           my_y = my_theta_1 + my_theta_2 * x
34           ax1.plot(x,my_y, color=line_colors[color_index],
35                   label='iter: {:.0f}'.format(i) + ' - ' +
36                   r'$\theta_1 = $' + '{:.2f}'.format(my_theta_1) +
37                   ' - ' +
38                   r'$\theta_2 = $' + '{:.2f}'.format(my_theta_2))
39 ax1.set_xlabel('x')
40 ax1.set_ylabel('y')
41 ax1.legend()
42 cost_function = np.array(cost_function)
43 iterations = range(t_final)
44 ax2.plot(iterations,cost_function, color='#C82127',
45         label='mean squared-error cost function Eq.4.29')
46 ax2.set_xlabel('Iteration')
47 ax2.set_ylabel('Cost Function Value')
48 ax2.legend()
49 fig.tight_layout()
```

Listing 7.2   A simple example of gradient descent in Python

The GD then optimizes the parameters of our model through an iterative approach:

$$\theta_1^{t+1} = \theta_1^t - \gamma D_{\theta_1}, \tag{7.32}$$

$$\theta_2^{t+1} = \theta_2^t - \gamma D_{\theta_2}, \tag{7.33}$$

where γ is an appropriate learning rate. Code listing 7.2 and Fig. 7.3 show how to develop the GD optimization described by Eqs. (7.28)–(7.32).

The stochastic gradient descent (SGD) algorithm (Bottou, 2012) simplifies the GD algorithm by estimating the gradient of $C(\boldsymbol{\theta})$ on the basis of a single, randomly

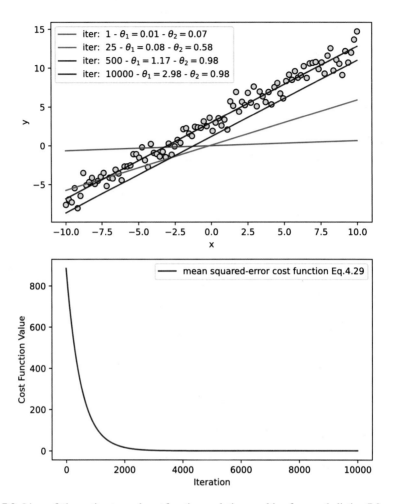

Fig. 7.3 Linear fitting estimates and cost function evolution resulting from code listing 7.2

picked example $\hat{f}_{\theta_t}(x_t)$:

$$\theta^{t+1} = \theta^t - \gamma \nabla C[y, \hat{f}_{\theta_t}(x_t)]. \tag{7.34}$$

The *SGDClassifier()* and *SGDRegressor()* in *sklearn.linear_model* implement a SGD in the field of classification and regression, respectively. Often, we use an approach that falls between GD and SGD by estimating the gradient using a small random portion of the training data set. This approach is called "mini-batch GD."

To summarize, GD always uses the entire learning data set. As opposed to GD, SGD and mini-batch GD compute the gradient from a single sample and a small portion of the training data set, respectively.

SGD and mini-batch GD work better than GD when numerous local maxima and minima occur. In this case, GD will probably stop at the first local minimum whereas SGD and mini-batch GD, being much noisier than GD, tend to explore neighboring areas of the gradient. Note that a pure SGD is significantly noisy, whereas mini-batch GD tends to average the computed gradient, resulting in more stable results than SGD. In ML, SGD and mini-batch GD see much more use than GD because the latter is too expensive computationally while providing only a minimum gain in accuracy for convex problems. For many local maxima and minima, SGD and mini-batch GD are also more accurate than GD because they can "jump" over local minima and hopefully find better solutions.

7.6 Ridge Regression

Ridge regression is a least squares method that shrinks the regression coefficients via a penalty on their size (Hastie et al., 2017). The regression starts with a labeled data set $(\mathbf{x}_i, y_i)$, where y_i are the labels and $\mathbf{x}_i = (x_{i1}, x_{i2}, \ldots, x_{ip})^T$ are the predictor variables (i.e., the inputs) (Hastie et al., 2017; Tibshirani, 1996).

The cost function in ridge regression can be expressed as (Hastie et al., 2017; Tibshirani, 1996)

$$C(\theta_0, \boldsymbol{\theta}) = \frac{1}{2n} \sum_{i=1}^{n} \left(y_i - \theta_0 - \sum_{j=1}^{p} x_{ij}\theta_j \right)^2 + \lambda \sum_{j=1}^{p} \theta_j^2, \tag{7.35}$$

where the parameter λ is called the "regularization penalty." The ridge regression performs the so-called L2-norm regularization by adding a penalty equivalent to the square of the magnitude of coefficients [i.e., the second term of Eq. (7.35)].

In the limiting case of $\lambda = 0$, the ridge regression reduces to an ordinary least squares regression. A correct choice of λ helps avoid overfitting issues. In contrast, underfitting becomes a problem for large λ.

7.7 Least Absolute Shrinkage and Selection Operator

The "least absolute shrinkage and selection operator," also known as the LASSO, is a method to solve linear problems by minimizing the residual sum of squares subject to the constraint that the sum of the absolute value of the coefficients must be less than a given constant (Tibshirani, 1996). The main characteristic of LASSO is its tendency to prefer solutions with fewer nonzero coefficients, thus reducing the number of parameters governing the predictor. The LASSO cost function can be expressed as (Tibshirani, 1996)

$$C(\theta_0, \boldsymbol{\theta}) = \frac{1}{2n} \sum_{i=1}^{n} \left(y_i - \theta_0 - \sum_{j=1}^{p} x_{ij}\theta_j \right)^2 + \lambda \sum_{j=1}^{p} |\theta_j|. \tag{7.36}$$

In contrast with ridge regression, the LASSO algorithm performs the so-called L1-norm regularization by adding a penalty equivalent to the sum of the absolute values of the coefficients [i.e., the second term of Eq. (7.36)].

Note that the LASSO reduces shrinkage and the dimensionality; in other words, it reduces the number of features of the solution, whereas ridge regression only shrinks (Hastie et al., 2017; Tibshirani, 1996).

7.8 Elastic Net

Elastic net (Zou & Hastie, 2005) is a linear regression model that performs both L1- and L2-norm regularization (Friedman et al., 2010):

$$C(\theta_0, \boldsymbol{\theta}) = \frac{1}{2n} \sum_{i=1}^{n} \left(y_i - \theta_0 - \sum_{j=1}^{p} x_{ij}\theta_j \right)^2 + \lambda \sum_{j=1}^{p} \left[\frac{1-\alpha}{2}\theta_j^2 + \alpha |\theta_j| \right].$$
$$\tag{7.37}$$

For $\alpha = 1$, elastic net is the same as the LASSO, whereas for $\alpha = 0$, elastic net approaches ridge regression. For $0 < \alpha < 1$, the penalty term [i.e., the second term of Eq. (7.37)] is between the L1- and L2-norm regularization.

7.9 Support Vector Machines

Support vector machines (SVMs) are a set of supervised ML algorithms that work remarkably well for classification (Cortes & Vapnik, 1995). The strength of SVMs relies on three features: (1) SVMs are efficient in high-dimensional spaces, (2) SVMs effectively model real-world problems, (3) SVMs perform well on data sets with many attributes, even if few cases are available to train the model (Cortes & Vapnik, 1995). SVMs numerically implement the following idea: inputs are nonlinearly mapped to a high-dimension feature space F (Cortes & Vapnik, 1995), and a linear decision surface is constructed in the space F (Cortes & Vapnik, 1995).

To start, consider a labeled training data set $(y_i, \boldsymbol{x_i})$, where $\boldsymbol{x_i}$ is p-dimensional [i.e., $\boldsymbol{x_i} = (x_{1i}, x_{2i}, \dots, x_{pi})$], with $i = 1, 2, \dots, n$, where n is the number of samples. Also, assume that the label $y_i = 1$ for the first class and $y_i = -1$ for the second class, defining a two-class classification problem (i.e., $y_i \in \{-1, 1\}$).

Now define a linear classifier based on the following linear-in-inputs discriminant function:

$$f(x) = \mathbf{w}^T \cdot x + b. \tag{7.38}$$

The decision boundary between the two classes (i.e., the regions classified as positive or negative) defined by Eq. (7.38) is a hyperplane.

The two classes are linearly separable if there exists a vector $\mathbf{w}$ and a scalar b such that

$$(\mathbf{w}^T \mathbf{x}_i + b) y_i \geq 1, \forall i = 1, 2, \ldots, n. \tag{7.39}$$

This means that we can correctly classify all samples. The definition of the optimal hyperplane follows as that which separates the training data set with a maximal margin

$$m(\mathbf{w}) = \frac{1}{\|\mathbf{w}\|}. \tag{7.40}$$

Finally, the maximum-margin classifier (i.e., the hard margin SVM) is the discriminant function that maximizes $m(\mathbf{w})$, which is equivalent to minimizing $\|\mathbf{w}\|^2$:

$$\min_{\mathbf{w},b} \frac{1}{2} \|\mathbf{w}\|^2 \tag{7.41}$$

subject to

$$(\mathbf{w}^T \mathbf{x}_i + b) y_i \geq 1, \forall i = 1, 2, \ldots, n. \tag{7.42}$$

The hard margin SVM requires the strong assumption of the linear separability of classes, which can be considered as an exception, not the rule. To allow errors [i.e., $\xi = (\xi_1, \xi_2, \ldots, \xi_n)$], we can introduce the concept of soft margin SVM:

$$\min_{\mathbf{w},b,\xi} \left[\frac{1}{2} \|\mathbf{w}\|^2 + C \sum_{i=1}^{n} \xi_i \right] \tag{7.43}$$

subject to

$$(\mathbf{w}^T \mathbf{x}_i + b) y_i \geq 1 - \xi_i, \xi_i \geq 0, \forall i = 1, 2, \ldots, n, \tag{7.44}$$

where $C > 0$ is a tunable parameter that controls the margin errors. The linear classifier defined by Eq. (7.38) can be generalized to nonlinear inputs by defining the discriminant function (Cortes & Vapnik, 1995)

$$f(\mathbf{x}) = \mathbf{w}^T \cdot \phi(\mathbf{x}) + b, \tag{7.45}$$

where $\phi(\mathbf{x})$ is a function that maps nonlinearly separable inputs $\mathbf{x}$ to a feature space F of higher dimension. If we express the weight vector $\mathbf{w}$ as a linear combination of the training examples (i.e., $\mathbf{w} = \sum_{i=1}^{n} \alpha_i \mathbf{x}_i$), it follows that, in feature space F, we have

$$f(\mathbf{x}) = \sum_{i=1}^{n} \alpha_i \phi(\mathbf{x_i})^T \phi(\mathbf{x}) + b. \tag{7.46}$$

The idea behind Eqs. (7.45) and (7.46) is to map a nonlinear classification function to a feature space F of higher dimensions, where the classification function is linear (Fig. 7.4). Defining a kernel function $K(\mathbf{x_i}, \mathbf{x})$ as

$$K(\mathbf{x_i}, \mathbf{x}) = \phi(\mathbf{x_i})^T \phi(\mathbf{x}), \tag{7.47}$$

we have

$$f(\mathbf{x}) = \sum_{i=1}^{n} \alpha_i K(\mathbf{x_i}, \mathbf{x}) + b. \tag{7.48}$$

When using the kernel function, we do not need to know or compute $\phi()$, which allows us to apply a linear transformation to the problem at higher dimensions. The scikit-learn implementation of SVMs [e.g., *SVC*() and *SVR*()] allows the use of linear, polynomial, sigmoid, and radial basis kernel functions [$K(\mathbf{x_i}, \mathbf{x})$, Table 7.1].

7.10 Supervised Nearest Neighbors

Supervised k-nearest neighbors is a ML algorithm that uses similarities such as distance functions (Bentley, 1975) to regress and classify. In detail, the k-nearest-neighbors method predicts numerical targets by using a metric that is typically the inverse-distance-weighted average of the k-nearest neighbors (Bentley, 1975). The weights can be uniform or calculated by a kernel function. The Euclidean distance metric is commonly used to measure the distance between two instances, although other metrics are available (see Table 7.2). Note that the Minkowski distance reduces to the Manhattan and Euclidean distances when $p = 1$ and 2, respectively. Bentley (1975) gives an extensive and detailed description of the k-nearest neighbors algorithm.

In scikit-learn, the *KNeighborsClassifier*() and *KNeighborsRegressor*() methods perform classification and regression, respectively, based on the k-nearest neighbors.

Fig. 7.4 Support vector machines redrawn from Sugiyama (2015)

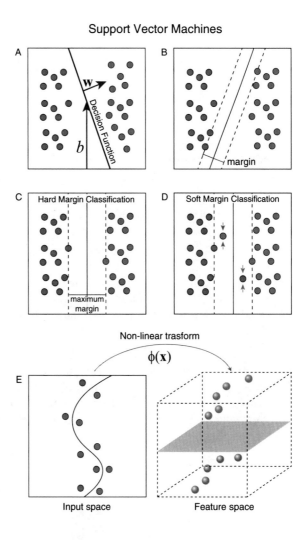

Table 7.1 Kernel functions in scikit-learn for the *SVC*() and *SVR*() methods

Kernel function	Equation	Identifier
Linear	$K(\mathbf{x_i}, \mathbf{x}) = (\mathbf{x_i} \cdot \mathbf{x'})$	kernel='linear'
Polynomial	$K(\mathbf{x_i}, \mathbf{x}) = (\mathbf{x_i} \cdot \mathbf{x'} + r)^d$	kernel='poly'
Sigmoid	$K(\mathbf{x_i}, \mathbf{x}) = tanh(\mathbf{x_i} \cdot \mathbf{x'} + r)$	kernel='sigmoid'
Radial basis function	$K(\mathbf{x_i}, \mathbf{x}) = exp(-\lambda \left\| \mathbf{x_i} - \mathbf{x'} \right\|^2)$	kernel='rbf'

Table 7.2 Selected distance metrics that can be used in supervised nearest neighbors and other ML algorithms

Distance	Identifier	Arguments	Equation
Euclidean	'euclidean'	None	$\sqrt{\sum_{j=1}^{D} \left|x_j - y_j\right|^2}$
Manhattan	'manhattan'	None	$\sum_{j=1}^{D} \left|x_j - y_j\right|$
Chebyshev	'chebyshev'	None	$\max \left|x_j - y_j\right|$
Minkowski	'minkowski'	$p, (w = 1)$	$\left(\sum_{j=1}^{D} w \left|x_j - y_j\right|^p\right)^{1/p}$

7.11 Trees-Based Methods

Decision Trees Before describing how decision trees work, let me introduce a few definitions highlighted in Fig. 7.5. A *root node* is the starting node of a decision tree and contains the entire data set involved in the process. A *parent node* is a node that is split into sub-nodes. A *child node* is a sub-node of a parent node. Finally, a *leaf* or *terminal node* are nodes that terminate the tree and that are not split to generate additional child nodes.

The decision tree algorithm (Breiman et al., 1984) and its modifications such as random forests and extra trees split the input space into sub-regions, which allow for regression and classification tasks (see Fig. 7.5) (Kubat, 2017). In detail, each node maps a region in the input space, which is further divided within the node into sub-regions by using splitting criteria. Therefore, the workflow of a decision

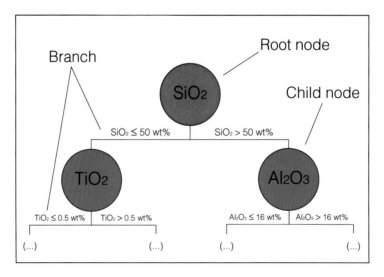

Fig. 7.5 The decision tree algorithm

tree consists of progressively splitting the input space by a sequence of decisions (i.e., splittings) into non-overlapping regions, with a one-to-one correspondence between leaf nodes and input regions (Kubat, 2017). Unfortunately, the decision tree algorithm, despite the appeal of the simplicity of its formulation, is often prone to overfitting and underfitting (cf. Sect. 3.5.5), making it less accurate than other predictors (Song & Lu, 2015).

To avoid overfitting and underfitting, more robust algorithms called ensemble predictors have been developed. Examples are random forest, gradient boosting, and extremely randomized tree methods. More details on the decision tree model are available from Breiman et al. (1984).

Random Forest The random forest algorithm (Breiman, 2001) is based on the "bagging" technique, which is a bootstrap aggregation technique that averages the prediction over a collection of bootstrap samples, thereby reducing the variance (Hastie et al., 2017). In detail, the random forest algorithm uses bagging to create multiple versions of a predictor (i.e., multiples trees), then evaluates the predictors to obtain an aggregated predictor (Hastie et al., 2017). Specifically, for a given training data set with sample size n, bagging produces k new training sets by uniformly sampling from the original training data set with replacement (i.e., by bootstrapping) (Hastie et al., 2017). Next, k decision trees are trained by using the k newly created training sets and are typically coupled by averaging for regression or majority voting for classification (Hastie et al., 2017). A detailed description of the random forest algorithm is available from Breiman (2001) and Hastie et al. (2017).

Extremely Randomized Trees The extremely randomized trees algorithm (Geurts et al., 2006) is similar to the random forest algorithm with two main differences: (i) it splits nodes by choosing fully random cut points and (ii) it uses the entire training sample rather than a bootstrapped replica to grow the trees (Geurts et al., 2006). The predictions of the trees are typically aggregated to yield the final prediction by majority vote in the classification and by arithmetic averaging in the regression (Geurts et al., 2006). A complete description of the extremely randomized trees algorithm is given by Geurts et al. (2006).

References

Bentley, J. L. (1975). Multidimensional binary search trees used for associative searching. *Communications of the ACM, 18*(9), 509–517. https://doi.org/10.1145/361002.361007

Bottou, L. (2012). Stochastic gradient descent tricks. In G. Montavon, G. B. Orr, & K.-R. Müller (Eds.), *Neural networks: Tricks of the trade* (2nd ed., pp. 421–436). Berlin: Springer. https://doi.org/10.1007/978-3-642-35289-8_25

Breiman, L. (2001). Random forests. *Machine Learning, 45*(1), 5–32. https://doi.org/10.1023/A:1010933404324

Breiman, L., Friedman, J. H., Olshen, R. A., & Stone, C. J. (1984). *Classification and regression trees*. Boca Raton: Chapman and Hall/CRC.

Cortes, C., & Vapnik, V. (1995). Support-vector networks. *Machine Learning, 20*(3), 273–297. https://doi.org/10.1007/BF00994018

Friedman, J., Hastie, T., & Tibshirani, R. (2010). Regularization paths for generalized linear models via coordinate descent. *Journal of Statistical Software, 33*(1), 1. https://doi.org/10.18637/jss.v033.i01

Geurts, P., Ernst, D., & Wehenkel, L. (2006). Extremely randomized trees. *Machine Learning, 63*(1), 3–42. https://doi.org/10.1007/S10994-006-6226-1

Goodfellow, I., Bengio, Y., & Courville, A. (2016). *Deep learning* (vol. 29). Cambridge: MIT Press.

Hastie, T., Tibshirani, R., & Friedman, J. (2017). *The elements of statistical learning* (2nd ed.). Berlin: Springer.

Kubat, M. (2017). *An introduction to machine learning.* Berlin: Springer. https://doi.org/10.1007/978-3-319-63913-0

Maronna, R. A., Martin, R. D., & Yohai, V. J. (2006). *Robust statistics: Theory and methods.* Hoboken: Wiley. https://doi.org/10.1002/0470010940

Song, Y. Y., & Lu, Y. (2015). Decision tree methods: Applications for classification and prediction. *Shanghai Archives of Psychiatry, 27*(2), 130. https://doi.org/10.11919/J.ISSN.1002-0829.215044

Sugiyama, M. (2015). *Introduction to statistical machine learning.* Amsterdam: Elsevier. https://doi.org/10.1016/C2014-0-01992-2

Tibshirani, R. (1996). Regression shrinkage and selection via the lasso. *Journal of the Royal Statistical Society: Series B (Methodological), 58*(1), 267–288. https://doi.org/10.1111/J.2517-6161.1996.TB02080.X

Zhang, H. (2004). The optimality of Naive Bayes. The Florida AI Research Society.

Zou, H., & Hastie, T. (2005). Regularization and variable selection via the elastic net. *Journal of the Royal Statistical Society: Series B (Statistical Methodology), 67*(2), 301–320. https://doi.org/10.1111/J.1467-9868.2005.00503.X

Chapter 8
Classification of Well Log Data Facies by Machine Learning

8.1 Motivation

Recognizing facies in wells through well-log data analysis is a common task in many geological fields such as trap reservoir characterization, sedimentology analysis, and depositional-environment interpretation (Hernandez-Martinez et al., 2013; Wood, 2021). I started conceiving this chapter when I discovered the FORCE 2020[1] ML competition (Bormann et al., 2020) and the SEG 2016[2] ML contest (M. Hall & Hall, 2017). In these two contests, students and early-career researchers attempt to identify lithofacies in a blind data set of well-log data (i.e., gamma-ray, resistivity, photoelectric effect, etc. . . .) by using a ML algorithm of their selection to be trained on a labeled data set made available to all competitors. The competitors of the 2016 edition were supported by a tutorial by Brendon Hall (B. Hall, 2016) and Hall and Hall (M. Hall & Hall, 2017). Also, Bestagini et al. (2017) described a strategy to achieve the final goal for the 2016 edition. Note that the starter notebook[3] of the FORCE 2020 ML competition contains all you need to begin: it shows how to import the training data set, inspect the imported data set, and start developing a model based on the random forest algorithm.

This chapter focuses on the FORCE 2020 Machine Learning competition by progressively developing a ML workflow (i.e., descriptive statistics, algorithm selection, model optimization, model training, and application of the model to the blind data set) and discussing each step to make everything as simple as possible.

---

[1] https://github.com/bolgebrygg/Force-2020-Machine-Learning-competition.

[2] https://github.com/seg/2016-ml-contest.

[3] https://bit.ly/force2020_ml_start.

M. Petrelli, *Machine Learning for Earth Sciences*, Springer Textbooks in Earth Sciences, Geography and Environment, https://doi.org/10.1007/978-3-031-35114-3_8

8.2 Inspection of the Data Sets and Pre-processing

For the FORCE 2020 Machine Learning competition,[4] a starter Jupyter Notebook has been made available on GitHub together with a labeled training dataset (i.e., the compressed train.zip file containing the single file train.csv) and two tests (i.e., leaderboard_test_features.csv and hidden_test.csv).[5] Nowadays, all three files are labeled, which means that they also contain the correct solution either in a column named FORCE_2020_LITHOFACIES_LITHOLOGY or in a separate file (Bormann et al., 2020). The above data set is provided by a NOLD 2.0[6] license and contains well-log data for more than 90 wells offshore of Norway (B. Hall, 2016; Bormann et al., 2020).

We start by importing the three data sets using pandas and looking at the spatial distribution of the wells under investigation (code listing 8.1; Fig. 8.1).

```
1  import pandas as pd
2  import matplotlib.pyplot as plt
3
4  data_sets = ['train.csv', 'hidden_test.csv', '
       leaderboard_test_features.csv']
5  labels = ['Train data', 'Hidden test data', 'Leaderboard test
       data']
6  colors = ['#BFD7EA','#0A3A54','#C82127']
7
8  fig, ax = plt.subplots()
9
10 for my_data_set, my_color, my_label in zip(data_sets, colors,
       labels):
11
12     my_data = pd.read_csv(my_data_set, sep=';')
13     my_Weels = my_data.drop_duplicates(subset=['WELL'])
14     my_Weels = my_Weels[['X_LOC', 'Y_LOC']].dropna() / 100000
15
16     ax.scatter(my_Weels['X_LOC'], my_Weels['Y_LOC'],
17                 label=my_label, s=80, color=my_color,
18                 edgecolor='k', alpha=0.8)
19
20 ax.set_xlabel('X_LOC')
21 ax.set_ylabel('Y_LOC')
22 ax.set_xlim(4,6)
23 ax.set_ylim(63,70)
24 ax.legend(ncol=3)
25 plt.tight_layout()
```

Listing 8.1 Spatial distribution of wells under investigation

---

[4] https://xeek.ai/challenges/force-well-logs/overview.

[5] https://bit.ly/force2020_ml_data.

[6] https://data.norge.no/nlod/en/2.0.

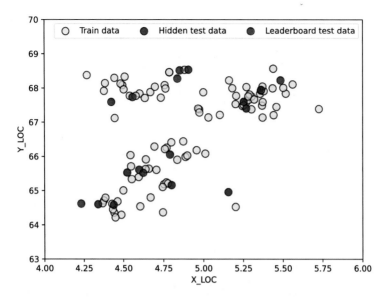

Fig. 8.1   Result of code listing 8.1. Spatial distribution of investigated wells

Figure 8.1 shows that the wells are distributed in three main clusters. As geologists, we would expect wells that are close together to have similar distributions of lithofacies. Therefore, well position could significantly impact the training of our ML model. Many strategies are available to include the spatial distribution of wells in a ML model; including X_LOC and Y_LOC as model features is the easiest strategy. More refined strategies may include a preliminary clustering of the spatial distribution of wells and a learning approach based on the result of the clustering. To develop a smart and simple workflow, we select the first option (i.e., simply including X_LOC and Y_LOC as model features).

Figure 8.2 shows the results of code listing 8.2, which reveal two main characteristics of the investigated data sets. The first characteristic relates to feature persistence. Many features such as Spectral Gamma Ray (SGR), Shear wave sonic log (DTS), Micro Resisitivity measurement (RMIC), and Average Rate of Penetration (ROPA) contain more than 60% missing values (see the upper panel of Fig. 8.2). Consequently, a strategy to deal with missing values is mandatory. Given our desire to maintain the simplicity of the ML workflow presented in the present chapter, only features containing less than 40% missing values are used. In addition, all missing values are replaced with the average of each feature. In statistics, the procedure of substituting missing values with other values is called "feature imputation" (Zou et al., 2015). In scikit-learn, *SimpleImputer*() and *IterativeImputer*() are useful for feature imputation (cf. Sect. 3.3.2).

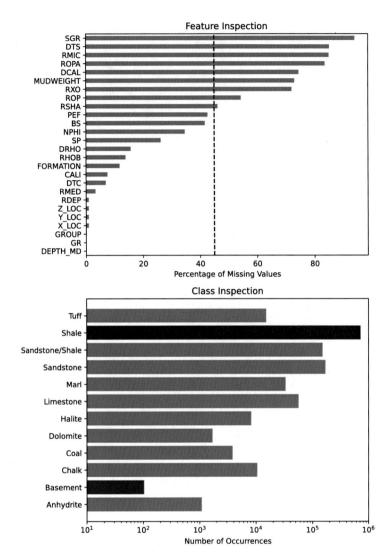

Fig. 8.2 Result of code listing 8.2. Inspect feature persistence and class balancing

```
1  import pandas as pd
2  import numpy as np
3  import matplotlib.pyplot as plt
4
5  lithology_keys = {30000: 'Sandstone',
6                    65030: 'Sandstone/Shale',
7                    65000: 'Shale',
8                    80000: 'Marl',
9                    74000: 'Dolomite',
```

```
10                         70000: 'Limestone',
11                         70032: 'Chalk',
12                         88000: 'Halite',
13                         86000: 'Anhydrite',
14                         99000: 'Tuff',
15                         90000: 'Coal',
16                         93000: 'Basement'}
17
18  train_data = pd.read_csv('train.csv', sep=';')
19
20  class_abundance = np.vectorize(lithology_keys.get)(
21      train_data['FORCE_2020_LITHOFACIES_LITHOLOGY'].values)
22  unique, counts = np.unique(class_abundance, return_counts=True)
23
24  my_colors = ['#0F7F8B'] * len(unique)
25  my_colors[np.argmax(counts)] = '#C82127'
26  my_colors[np.argmin(counts)] = '#0A3A54'
27
28  fig, (ax1, ax2) = plt.subplots(2,1, figsize=(7,14))
29
30  ax2.barh(unique,counts, color=my_colors)
31  ax2.set_xscale("log")
32  ax2.set_xlim(1e1,1e6)
33  ax2.set_xlabel('Number of Occurrences')
34  ax2.set_title('Class Inspection')
35
36  Feature_presence = train_data.isna().sum()/train_data.shape
        [0]*100
37
38  Feature_presence =Feature_presence.drop(
39              labels=['FORCE_2020_LITHOFACIES_LITHOLOGY',
40                      'FORCE_2020_LITHOFACIES_CONFIDENCE', 'WELL
        '])
41
42  Feature_presence.sort_values().plot.barh(color='#0F7F8B',ax=ax1)
43  ax1.axvline(40, color='#C82127', linestyle='--')
44  ax1.set_xlabel('Percentage of Missing Values')
45  ax1.set_title('Feature Inspection')
46
47  plt.tight_layout()
```

Listing 8.2 Inspect feature persistence and class balancing

The second key characteristic of the investigated data set appears clearly upon observing the class distribution (see lower panel of Fig. 8.2): the training data set is highly imbalanced, with some classes exceeding 10^5 occurrences and others such as Anhydrite and Basement occurring only 10^3 or 10^2 times, respectively. A strategy to account for the imbalance of training data set is thus also mandatory.

Some ML algorithms, such as those discussed in the present chapter, try to account for imbalance in their training data set by tuning their hyperparameters. More refined strategies may involve (1) under-sampling majority classes, (2) over-

sampling minority classes, (3) combining over- and under-sampling methods, and
(4) creating ensemble balanced sets (Lemaître et al., 2017).

```
 1  import numpy as np
 2
 3  fig = plt.figure(figsize=(8,4))
 4
 5  train_data['log_RDEP'] = np.log10(train_data['RDEP'])
 6
 7  to_be_plotted = ['RDEP', 'log_RDEP']
 8
 9  for index, my_feature in enumerate(to_be_plotted):
10      ax = fig.add_subplot(1,2,index+1)
11      min_val = np.nanpercentile(train_data[my_feature],1)
12      max_val = np.nanpercentile(train_data[my_feature],99)
13      my_bins = np.linspace(min_val,max_val,30)
14      ax.hist(train_data[my_feature], bins=my_bins,
15              density = True,  color='#BFD7EA',
16              edgecolor='k')
17      ax.set_ylabel('Probability Density')
18      ax.set_xlabel(my_feature)
19
20  plt.tight_layout()
```

Listing 8.3 Log-transformation of selected features

The histogram distribution of some features (code listing 8.3 and Fig. 8.3)
shows that they are highly skewed, which could be a problem for some ML
algorithms (e.g., those assuming a normal distribution for the investigated features).
Consequently, we apply a log-transformation to selected features to reduce the
skewness (see Fig. 8.3, right panel).

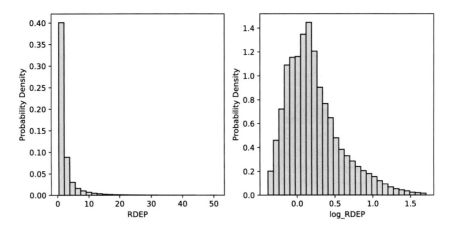

Fig. 8.3 Result of code listing 8.3. Log-transformation of selected features

As discussed in Chap. 3.3, the goal of data augmentation is to improve the generalizability of ML models by increasing the amount of information in their data sets. This approach consists of adding modified copies (e.g., flipper or rotated images in the case of image classification) of the available data or combining the existing features to generate new features. For example, Bestagini et al. (2017) suggest three approaches for data augmentation: quadratically expanding the feature vector, considering second-order interaction terms, and defining an augmented gradient feature vector. In an attempt to partially mimic the data augmentation strategy proposed by Bestagini et al. (2017), we report a code listing to calculate the augmented gradient feature vector (code listing 8.4).

```
1  def calculate_delta(dataFrame):
2      delta_features = ['CALI', 'log_RMED', 'log_RDEP', 'RHOB', '
       DTC', 'DRHO', 'log_GR' , 'NPHI', 'log_PEF', 'SP']
3      wells = dataFrame['WELL'].unique()
4      for my_feature in delta_features:
5          values = []
6          for well in wells:
7              col_values = dataFrame[dataFrame['WELL'] == well][
       my_feature].values
8              col_values_ = np.array([col_values[0]]+list(
       col_values[:-1]))
9              delta_col_values = col_values-col_values_
10             values = values + list(delta_col_values)
11         dataFrame['Delta_' + my_feature] = values
12     return dataFrame
```

Listing 8.4 Function to calculate the augmented gradient feature vector

To summarize, our pre-processing strategy starts with (i) selecting the features characterized by fewer than 40% missing values, (ii) replacing missing values with the average of each feature within each data set, (iii) applying a log-transformation of the features with highly skewed distributions, and (iv) augmenting the data. Steps (i)–(iv) are implemented in a series of functions (see code listing 8.5) and are combined in a pandas *pipe()* chain to automate pre-processing (code listing 8.6). Also, the *pre_processing_pipeline()* function (code listing 8.6) stores the imported .csv files in a single HDF5 file (hierarchical data format version 5). As introduced in Sect. 3.3, HDF5 is a high-performance library to manage, process, and store heterogeneous data. All data sets of interest are stored in HDF5 files as pandas DataFrames, ready for fast reading and writing. At lines 3–6, the function checks that the output file exists. If so, the function removes the existing file. At line 16, the function appends each processed data set to a newly created file.

```
 1  import os
 2  import pandas as pd
 3  import numpy as np
 4
 5  def replace_inf(dataFrame):
 6      to_be_replaced = [np.inf,-np.inf]
 7      for replace_me in to_be_replaced:
 8          dataFrame = dataFrame.replace(replace_me, np.nan)
 9      return dataFrame
10
11  def log_transform(dataFrame):
12      log_features = ['RDEP','RMED','PEF','GR']
13      for my_feature in log_features:
14          dataFrame.loc[dataFrame[my_feature] < 0, my_feature] =
        dataFrame[dataFrame[my_feature] > 0].RDEP.min()
15          dataFrame['log_'+ my_feature] = np.log10(dataFrame[
        my_feature])
16      return dataFrame
17
18  def calculate_delta(dataFrame):
19      delta_features = ['CALI', 'log_RMED', 'log_RDEP', 'RHOB',
20                        'DTC', 'DRHO', 'log_GR' , 'NPHI',
21                        'log_PEF', 'SP']
22      wells = dataFrame['WELL'].unique()
23      for my_feature in delta_features:
24          values = []
25          for well in wells:
26              my_val = dataFrame[dataFrame['WELL'] == well][
        my_feature].values
27              my_val_ = np.array([my_val[0]] +
28                               list(my_val[:-1]))
29              delta_my_val = my_val-my_val_
30              values = values + list(delta_my_val)
31          dataFrame['Delta_' + my_feature] = values
32      return dataFrame
33
34  def feature_selection(dataFrame):
35      features = ['CALI', 'Delta_CALI',  'log_RMED',
36                  'Delta_log_RMED', 'log_RDEP',
37                  'Delta_log_RDEP', 'RHOB', 'Delta_RHOB',
38                  'SP', 'Delta_SP', 'DTC', 'Delta_DTC',
39                  'DRHO', 'Delta_DRHO', 'log_GR', 'Delta_log_GR',
40                  'NPHI', 'Delta_NPHI', 'log_PEF', 'Delta_log_PEF']
41      dataFrame = dataFrame[features]
42      return dataFrame
```

Listing 8.5 Defining the pre-processing functions

Figure 8.4 shows the results of code listing 8.7 and describes most of the numerical features to be used during training. These features are derived by applying the pre-processing strategy developed in code listings 8.5 and 8.6. All the features reported in Fig. 8.4 are numerically continuous. However, the investigated data sets also contain categorical features such as GROUP and FORMATIONS.

Most ML algorithms support the use of categorical features only after encoding to their numerical counterparts. Code listing 8.8 shows the *pipe*() chain of code listing 8.6 [i.e., *pre_processing_pipeline*()], with the addition of a categorical encoder to allow FORMATIONS to be investigated by a ML algorithm. We use the *OrdinalEncoder*() method from scikit-learn. Also, code listing 8.8 presents a modified version of the function *feature_selection*() to include the encoded feature FORMATION.

```python
def pre_processing_pipeline(input_files, out_file):

    try:
        os.remove(out_file)
    except OSError:
        pass

    for ix, my_file in enumerate(input_files):
        my_dataset = pd.read_csv(my_file, sep=';')

        try:
            my_dataset['FORCE_2020_LITHOFACIES_LITHOLOGY'].to_hdf(
                out_file, key=my_file[:-4] + '_target')
        except:
            my_target = pd.read_csv('leaderboard_test_target.csv', sep=';')
            my_target['FORCE_2020_LITHOFACIES_LITHOLOGY'].to_hdf(
                out_file, key=my_file[:-4] + '_target')

        if ix==0:
            # Fitting the categorical encoders
            my_encoder = OrdinalEncoder()
            my_encoder.set_params(handle_unknown='use_encoded_value',
                                  unknown_value=-1,
                                  encoded_missing_value=-1).fit(
                                      my_dataset[['FORMATION']])

        my_dataset = (my_dataset.
                         pipe(replace_inf).
                         pipe(log_transform).
                         pipe(calculate_delta).
                         pipe(feature_selection))
        my_dataset.to_hdf(out_file, key=my_file[:-4])

        my_dataset.to_hdf(out_file, key= my_file[:-4])
```

```
35
36 my_files = ['train.csv', 'leaderboard_test_features.csv', '
      hidden_test.csv']
37
38 pre_processing_pipeline(input_files=my_files, out_file='ml_data.
      h5')
```

Listing 8.6 Combining the pre-processing functions in a pandas *pipe*()

```
1 import pandas as pd
2 import numpy as np
3 import matplotlib.pyplot as plt
4
5 train_data = pd.read_hdf('ml_data.h5', 'train')
6 test_data = pd.read_hdf('ml_data.h5', 'leaderboard_test_features'
      )
7
8 show_axes = [1,5,9,13,17]
9 fig = plt.figure(figsize=(9, 15))
10
11 for i, my_feature in enumerate(train_data.columns[0:20], start=1)
      :
12     ax = fig.add_subplot(5,4,i)
13     min_val = np.nanpercentile(train_data[my_feature],1)
14     max_val = np.nanpercentile(train_data[my_feature],99)
15     my_bins = np.linspace(min_val,max_val,30)
16     ax.hist(train_data[my_feature], bins=my_bins, density = True,
17             histtype='step', color='#0A3A54')
18     ax.hist(test_data[my_feature], bins=my_bins, density = True,
19             histtype='step', color='#C82127', linestyle='--')
20     ax.set_xlabel(my_feature)
21     ymin, ymax = ax.get_ylim()
22     if ymax >=10:
23         ax.set_yticks(np.round(np.linspace(ymin, ymax, 4), 0))
24     elif ((ymax<10)&(ymax>1)):
25         ax.set_yticks(np.round(np.linspace(ymin, ymax, 4), 1))
26     else:
27         ax.set_yticks(np.round(np.linspace(ymin, ymax, 4), 2))
28
29     if i in show_axes:
30         ax.set_ylabel('Probability Density')
31
32 plt.tight_layout()
33 fig.align_ylabels()
```

Listing 8.7 Descriptive statistics

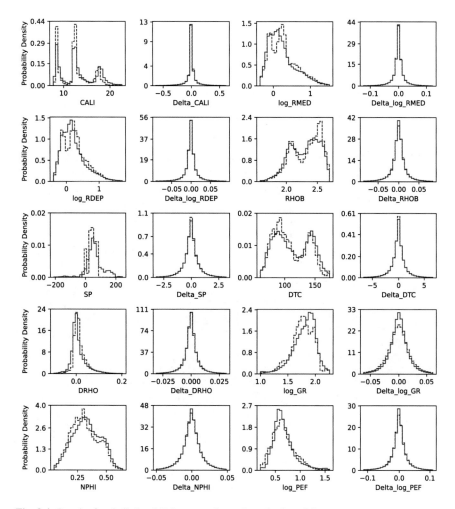

Fig. 8.4 Result of code listing 8.7. Log-transformation of selected features

```
1 import os
2 import pandas as pd
3 import numpy as np
4 from sklearn.preprocessing import OrdinalEncoder
5 from sklearn.impute import SimpleImputer
6
7 def replace_inf(dataFrame):
8     to_be_replaced = [np.inf,-np.inf]
9     for replace_me in to_be_replaced:
10         dataFrame = dataFrame.replace(replace_me, np.nan)
11     return dataFrame
12
13 def log_transform(dataFrame):
```

```
14      log_features = ['RDEP','RMED','PEF','GR']
15      for my_feature in log_features:
16          dataFrame.loc[dataFrame[my_feature] < 0, my_feature] =
        dataFrame[
17              dataFrame[my_feature] > 0].RDEP.min()
18          dataFrame['log_'+ my_feature] = np.log10(dataFrame[
        my_feature])
19      return dataFrame
20
21  def calculate_delta(dataFrame):
22      delta_features = ['CALI', 'log_RMED', 'log_RDEP', 'RHOB',
23                       'DTC', 'DRHO', 'log_GR' , 'NPHI',
24                       'log_PEF', 'SP', 'BS']
25      wells = dataFrame['WELL'].unique()
26      for my_feature in delta_features:
27          values = []
28          for well in wells:
29              my_val = dataFrame[dataFrame['WELL'] == well][
        my_feature].values
30              my_val_ = np.array([my_val[0]] +
31                                  list(my_val[:-1]))
32              delta_my_val = my_val-my_val_
33              values = values + list(delta_my_val)
34          dataFrame['Delta_' + my_feature] = values
35      return dataFrame
36
37  def categorical_encoder(dataFrame, my_encoder, cols):
38      dataFrame[cols] =  my_encoder.transform(dataFrame[cols])
39      return dataFrame
40
41  def feature_selection(dataFrame):
42      features = ['CALI', 'Delta_CALI', 'log_RMED', '
        Delta_log_RMED',
43                  'log_RDEP','Delta_log_RDEP', 'RHOB', 'Delta_RHOB'
        ,
44                  'SP', 'Delta_SP', 'DTC', 'Delta_DTC', 'DRHO', '
        Delta_DRHO',
45                  'log_GR', 'Delta_log_GR', 'NPHI', 'Delta_NPHI',
46                  'log_PEF', 'Delta_log_PEF', 'BS', 'Delta_BS',
47                  'FORMATION', 'X_LOC','Y_LOC', 'DEPTH_MD']
48      dataFrame = dataFrame[features]
49      return dataFrame
50
51  def pre_processing_pipeline(input_files, out_file):
52
53      try:
54          os.remove(out_file)
55      except OSError:
56          pass
57
58      for ix, my_file in enumerate(input_files):
59          my_dataset = pd.read_csv(my_file, sep=';')
60
61          try:
```

```
62          my_dataset['FORCE_2020_LITHOFACIES_LITHOLOGY'].to_hdf
        (
63              out_file, key=my_file[:-4] + '_target')
64      except:
65          my_target = pd.read_csv('leaderboard_test_target.csv'
    , sep=';')
66          my_target['FORCE_2020_LITHOFACIES_LITHOLOGY'].to_hdf(
67              out_file, key=my_file[:-4] + '_target')
68
69      if ix==0:
70          # Fitting the categorical encoders
71          my_encoder = OrdinalEncoder()
72          my_encoder.set_params(handle_unknown='
    use_encoded_value',
73                                unknown_value=-1,
74                                encoded_missing_value=-1).fit(
75                                    my_dataset[['FORMATION']])
76
77      my_dataset = (my_dataset.
78                      pipe(replace_inf).
79                      pipe(log_transform).
80                      pipe(calculate_delta).
81                      pipe(categorical_encoder,
82                          my_encoder=my_encoder, cols=['
    FORMATION']).
83                      pipe(feature_selection))
84      my_dataset.to_hdf(out_file, key=my_file[:-4])
85
86      imputer = SimpleImputer(missing_values=np.nan, strategy='
    mean')
87      imputer = imputer.fit(my_dataset[my_dataset.columns])
88      my_dataset[my_dataset.columns] = imputer.transform(
89          my_dataset[my_dataset.columns])
90      my_dataset.to_hdf(out_file, key= my_file[:-4])
91
92  my_files = ['train.csv', 'leaderboard_test_features.csv', '
        hidden_test.csv']
93
94  pre_processing_pipeline(input_files=my_files, out_file='ml_data.
        h5')
```

Listing 8.8 Pre-processing *pipe*() chain, including the categorical features

8.3   Model Selection and Training

After data pre-processing, the next fundamental steps are model selection, optimization, and training. Recall that we are dealing with a classification problem, so we
select from among supervised algorithms. In the following, we test the extremely
randomized trees algorithm [i.e., *ExtraTreesClassifier*()] in scikit-learn. Selecting
ExtraTreesClassifier() is an arbitrary choice and the reader is invited to explore
different ML methods, such as support vector machines.

In our specific case, the *ExtraTreesClassifier*() depends on many hyperparameters such as the number of trees, the number of investigated features, and the splitting criterion.

```
 1 import pandas as pd
 2 from sklearn.ensemble import ExtraTreesClassifier
 3 from sklearn.model_selection import train_test_split
 4 from sklearn.model_selection import GridSearchCV
 5 import joblib as jb
 6 from sklearn.preprocessing import StandardScaler
 7
 8 X = pd.read_hdf('ml_data.h5', 'train').values
 9 y = pd.read_hdf('ml_data.h5', 'train_target').values
10
11 X_train, X_test, y_train, y_test = train_test_split(
12     X, y, test_size=0.2, random_state=10, stratify=y)
13
14 scaler = StandardScaler()
15 X_train = scaler.fit_transform(X_train)
16
17 param_grid = {
18     'criterion': ['entropy', 'gini'],
19     'min_samples_split': [2, 5, 8, 10],
20     'max_features': ['sqrt', 'log2', None],
21     'class_weight': ['balanced', None]
22     }
23
24 classifier = ExtraTreesClassifier(n_estimators=250,
25                                   n_jobs=-1)
26
27 CV_rfc = GridSearchCV(estimator=classifier, param_grid=param_grid
        , cv= 3, verbose=10)
28 CV_rfc.fit(X_train, y_train)
29
30 jb.dump(CV_rfc, 'ETC_grid_search_results_rev_2.pkl')
```

Listing 8.9 Grid search using *GridSearchCV*()

All these hyperparameters may assume different values, which may positively or negatively affect the classification capability of the model. The easiest way to find the best combination of the investigated hyperparameters is to do a grid search, which consists of defining the most relevant values for each hyperparameter and then training and evaluating a model for each possible combination. Table 8.1 lists the hyperparameters selected for the grid search. The *GridSearchCV*() method in scikit-learn is used to do a grid search in Python (code listing 8.9). After importing all required libraries (lines 1–6), the pre-processed training data set is imported (line 8) with labels (line 9). Next, the training data set is split into two, with one part (i.e., *X_train*) for the training and validation within the grid search, and another part (i.e., *X_test*), never involved in the training, to test the results obtained during

Table 8.1 Hyperparameters used in the grid search to optimize the *ExtraTreesClassifier()* algorithm. Descriptions are from the scikit-learn documentation

Parameter	Description[a]	Values
Criterion	The function to measure the quality of a split.	['entropy', 'gini']
min_samples_split	The minimum number of samples required to split an internal node	[2, 5, 8, 10]
max_features	The number of features to consider when looking for the best split	['sqrt', 'log2', None]
class_weight	Weights associated with classes	['balanced', None]

[a] https://scikit-learn.org/stable/modules/generated/sklearn.ensemble.ExtraTreesClassifier.html

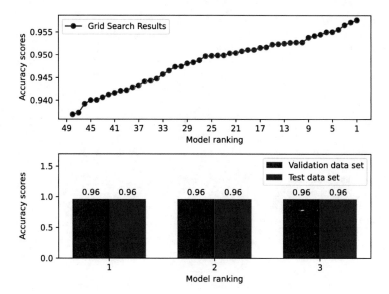

Fig. 8.5 Result of code listing 8.10

the grid search and for further testing against potential issues such as overfitting. The next step (lines 14 and 15) consists of scaling the data set involved in the grid search to zero mean and unit variance (cf. paragraph 3.3.5). Lines 17–22 define the set of parameters involved in the grid search. The combination of the selected hyperparameters results in a grid of 48 models, each repeated three times (cv = 3 at line 27) through cross validation (see Sect. 3.5.2) for a total of 144 attempts.

Running the code listing 8.9 on my MacBook pro, equipped wit a 2.3 GHz Quad-Core Intel™ Core i7 and 32 GB of RAM, takes about 8 hours. The top panel of Fig. 8.5 displays the accuracy scores of all 48 models, ordered by their ranking (code listing 8.10), and highlights that the best-performing models produce accuracy scores greater than 0.95. Such a strong performance may suggest that we are overfitting the training data set, so, as a first step, we use the three best-performing models (code listing 8.10) on the test data set (i.e., *X_test*). The bottom panel of Fig. 8.5 shows that the accuracy scores for *X_test* are of the same order of magnitude as those resulting from the grid search cross validation (i.e., $\approx$0.96), which does not support the idea of strong overfitting.

```python
from joblib import load
import numpy as np
import matplotlib.pyplot as plt
import pandas as pd
from sklearn.ensemble import ExtraTreesClassifier
from sklearn.model_selection import train_test_split
from sklearn.preprocessing import StandardScaler

CV_rfc = load('ETC_grid_search_results_rev_2.pkl')

my_results = pd.DataFrame.from_dict(CV_rfc.cv_results_)
my_results = my_results.sort_values(by=['rank_test_score'])

# Plot the results of the GridSearch
fig = plt.figure()
ax1 = fig.add_subplot(2,1,1)
ax1.plot(my_results['rank_test_score'], my_results['
    mean_test_score'], marker='o',
         markeredgecolor='#0A3A54', markerfacecolor='#C82127',
    color='#0A3A54',
         label='Grid Search Results')
ax1.set_xticks(np.arange(1,50,4))
ax1.invert_xaxis()
ax1.set_xlabel('Model ranking')
ax1.set_ylabel('Accuracy scores')
ax1.legend()

# Selecting the best three performing models
my_results = my_results[my_results['mean_test_score']>0.956]

# Load and scaling
X = pd.read_hdf('ml_data.h5', 'train').values
y = pd.read_hdf('ml_data.h5', 'train_target').values

X_train, X_test, y_train, y_test = train_test_split(
    X, y, test_size=0.2, random_state=10, stratify=y)

scaler = StandardScaler()
X_train = scaler.fit_transform(X_train)
X_test = scaler.transform(X_test)

leaderboard_test_features = pd.read_hdf('ml_data.h5', '
    leaderboard_test_features').values
hidden_test = pd.read_hdf('ml_data.h5', 'hidden_test').values

leaderboard_test_features_scaled = scaler.transform(
    leaderboard_test_features)
```

```
44 hidden_test_scaled = scaler.transform(hidden_test)
45
46 # Apply the three best performing model on the test dataset and
       on the unknowns
47 leaderboard_test_res = {}
48 hidden_test_res   = {}
49 test_score   = []
50 rank_model   = []
51 for index, row in my_results.iterrows():
52     classifier = ExtraTreesClassifier(n_estimators=250, n_jobs=8,
         random_state=64, **row['params'])
53     classifier.fit(X_train, y_train)
54     my_score = classifier.score(X_test,y_test)
55     test_score.append(my_score)
56     rank_model.append(row['rank_test_score'])
57
58     my_leaderboard_test_res = classifier.predict(
       leaderboard_test_features_scaled)
59     my_hidden_test_res = classifier.predict(hidden_test_scaled)
60     leaderboard_test_res['model_ranked_' + str(row['
       rank_test_score'])] = my_leaderboard_test_res
61     hidden_test_res['model_ranked_' + str(row['rank_test_score'])
       ] = my_hidden_test_res
62
63 leaderboard_test_res_pd = pd.DataFrame.from_dict(
       leaderboard_test_res)
64 hidden_test_res_pd = pd.DataFrame.from_dict(hidden_test_res)
65 leaderboard_test_res_pd.to_hdf('ml_data.h5', key= '
       leaderboard_test_res')
66 hidden_test_res_pd.to_hdf('ml_data.h5', key= 'hidden_test_res')
67
68 # plot the resultson the test dataset
69 ax2 = fig.add_subplot(2,1,2)
70 labels = my_results['rank_test_score']
71 validation_res = np.around(my_results['mean_test_score'], 2)
72 test_res = np.around(np.array(test_score),2)
73 x = np.arange(len(labels))
74 width = 0.35
75 rects1 = ax2.bar(x - width/2, validation_res, width, label='
       Validation data set', color='#C82127')
76 rects2 = ax2.bar(x + width/2, test_res, width, label='Test data
       set', color='#0A3A54')
77 ax2.set_ylabel('Accuracy scores')
78 ax2.set_xlabel('Model ranking')
79 ax2.set_ylim(0,1.7)
80 ax2.set_xticks(x, labels)
81 ax2.legend()
82 ax2.bar_label(rects1, padding=3)
83 ax2.bar_label(rects2, padding=3)
84 fig.align_ylabels()
85 fig.tight_layout()
```

Listing 8.10 Applying the three best-performing models on the test data set and on unknown samples

```
 1 import numpy as np
 2 import matplotlib.pyplot as plt
 3 from sklearn.metrics import accuracy_score
 4 import pandas as pd
 5
 6 leaderboard_test_res= pd.read_hdf('ml_data.h5', '
      leaderboard_test_res')
 7 hidden_test_res = pd.read_hdf('ml_data.h5', 'hidden_test_res')
 8
 9 leaderboard_test_target = pd.read_hdf('ml_data.h5', '
      leaderboard_test_features_target').values
10 hidden_test_target = pd.read_hdf('ml_data.h5', '
      hidden_test_target').values
11
12 leaderboard_accuracy_scores = []
13 hidden_accuracy_scores = []
14
15 for (leaderboard_column, leaderboard_data), (hidden_column,
      hidden_data) in zip(leaderboard_test_res.iteritems(),
      hidden_test_res.iteritems()):
16
17     leaderboard_accuracy_scores.append(np.around(accuracy_score(
        leaderboard_data, leaderboard_test_target),2))
18     hidden_accuracy_scores.append(np.around(accuracy_score(
        hidden_data, hidden_test_target),2))
19
20
21 # plot the resultson the test dataset
22 plt, ax1 = plt.subplots()
23 labels = leaderboard_test_res.columns
24 x = np.arange(len(labels))
25 width = 0.35
26 rects1 = ax1.bar(x - width/2, leaderboard_accuracy_scores, width,
       label='Leaderboard test data set', color='#C82127')
27 rects2 = ax1.bar(x + width/2, hidden_accuracy_scores, width,
       label='Hidden test est data set', color='#0A3A54')
28 ax1.set_ylabel('Accuracy scores')
29 #ax1.set_xlabel('Model ranking')
30 ax1.set_ylim(0,1.1)
31 ax1.set_xticks(x, labels)
32 ax1.legend()
33 ax1.bar_label(rects1, padding=3)
34 ax1.bar_label(rects2, padding=3)
```

Listing 8.11 Plotting the results obtained from the Leaderboard and the hidden test data sets

The three best-performing models were also run to predict the unknown samples (i.e., the leaderboard and the hidden test data sets). The accuracy scores (Fig. 8.6) for the leaderboard and the hidden test data sets (i.e., from 0.79 to 0.81) highlight that our ML models still perform with satisfaction on independent test data sets, so we move to the next section where we check the models against the evaluation criteria of the FORCE2020 challenge.

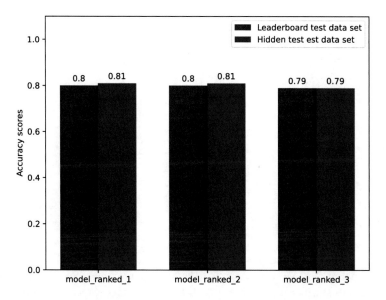

Fig. 8.6 Result of code listing 8.11

8.4 Final Evaluation

To evaluate the goodness of each model, the FORCE2020 challenge used a custom scoring strategy based on a penalty matrix (code listing 8.12).

```
import numpy as np

A = np.load('penalty_matrix.npy')
def score(y_true, y_pred):
    S = 0.0
    y_true = y_true.astype(int)
    y_pred = y_pred.astype(int)
    for i in range(0, y_true.shape[0]):
        S -= A[y_true[i], y_pred[i]]
    return S/y_true.shape[0]
```

Listing 8.12 Custom scoring function

In code listing 8.12, *y_true* and *y_pred* are the expected (i.e., correct) and predicted values, respectively, converted into integer indexes ranging from 0 to 11, as reported in Table 8.2.

The main objective of the FORCE2020 scoring strategy is to penalize errors made on easy-to-recognize lithologies more strongly than those made on difficult-to-recognize lithologies. To achieve this goal, the *score*() function weights each

Table 8.2 Connecting the labeling in the target files with lithofacie names and the indexing of the score function

Label	Lithofacie	Index
30000	'Sandstone'	0
65030	'Sandstone/Shale'	1
65000	'Shale'	2
80000	'Marl'	3
74000	'Dolomite'	4
70000	'Limestone'	5
70032	'Chalk'	6
88000	'Halite'	7
86000	'Anhydrite'	8
99000	'Tuff'	9
90000	'Coal'	10
93000	'Basement'	11

true-value–predicted-value pair by using the penalty matrix (code listing 8.13) reported in Fig. 8.7. More specifically, the *score*() function returns the value of the penalty matrix corresponding to each true-value–predicted-value pair (e.g., 4 if you confuse a Halite for a Sandstone; see Fig. 8.7). Next, the function sums all the scoring values and then calculates an "average" score by dividing the resulting value by the number of predictions.

```
1  import numpy as np
2  import matplotlib.pyplot as plt
3  import seaborn as sns
4
5  A = np.load('penalty_matrix.npy')
6
7  my_labels = ['Sandstone','Sandstone/Shale','Shale','Marl', '
       Dolomite',
8               'Limestone','Chalk','Halite','Anhydrite','Tuff','
       Coal','Basement']
9
10 fig, ax = plt.subplots(figsize=(15, 12))
11 ax.imshow(A)
12 ax = sns.heatmap(A, annot=True, xticklabels = my_labels,
       yticklabels = my_labels)
13 fig.tight_layout()
```

Listing 8.13 Penalty matrix

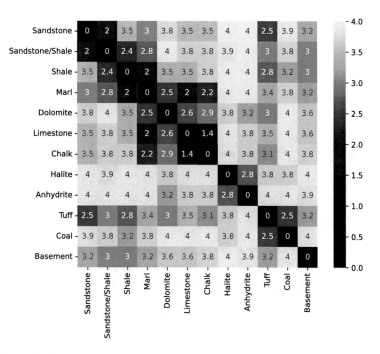

Fig. 8.7 Result of code listing 8.13

```
1  import numpy as np
2  import pandas as pd
3
4  A = np.load('penalty_matrix.npy')
5  def score(y_true, y_pred):
6      S = 0.0
7      y_true = y_true.astype(int)
8      y_pred = y_pred.astype(int)
9      for i in range(0, y_true.shape[0]):
10         S -= A[y_true[i], y_pred[i]]
11     return S/y_true.shape[0]
12
13 target = np.full(1000, 5) # Limestone
14 predicted = np.full(1000, 5)  # Limestone
15 print("Case 1: " + str(score(target, predicted)))
16
17 predicted = np.full(1000, 6)  # Chalk
18 print("Case 2: " + str(score(target, predicted)))
19
20 predicted = np.full(1000, 7) # Halite
21 print("Case 3: " + str(score(target, predicted)))
22
23 hidden_test_target = pd.read_hdf('ml_data.h5',
24                                   'hidden_test_target').values
25 predicted = np.random.randint(0, high=12,
```

```
26                                   size=1000) # Random predictions
27 print("Case 4: " + str(score(target, predicted)))
28
29 ''' Output:
30
31 Case 1: 0.0
32 Case 2: -1.375
33 Case 3: -4.0
34 Case 4: -3.04625
35
36 '''
```

Listing 8.14 Custom scoring function

Based on Fig. 8.7 and code listing 8.12, we can argue that a correct prediction
contributes zero to the score.

Therefore, if you correctly guess all the predictions, the score function returns
zero (see code listing 8.14, Case 1). In contrast, systematically predicting chalk
on a data set of limestone samples returns −1.375 (code listing 8.14, Case 2).
Systematically predicting halite on a data set of limestone samples returns −4.0
(code listing 8.14, Case 3), which is much more penalized than Case 2. Finally,
considering the hidden test data set, a dummy model providing random predictions
produces a score close to −3 (code listing 8.14, Case 4).

Figure 8.8 shows the result of applying the scoring strategy described above
to the leaderboard and hidden test data sets. Despite their simplicity, the two

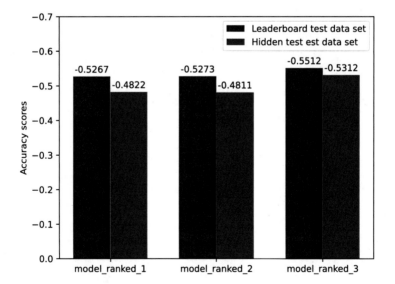

Fig. 8.8 Result of code listing 8.15

best-performing models produced by the grid search implemented in code listing 8.9, shows similar performances (i.e., > -0.50) of top-ranked models in the FORCE2020 challenge.[7]

```python
import matplotlib.pyplot as plt
import pandas as pd
import numpy as np

A = np.load('penalty_matrix.npy')
def score(y_true, y_pred):
    S = 0.0
    y_true = y_true.astype(int)
    y_pred = y_pred.astype(int)
    for i in range(0, y_true.shape[0]):
        S -= A[y_true[i], y_pred[i]]
    return S/y_true.shape[0]

lithology_numbers = {30000: 0, 65030: 1, 65000: 2, 80000: 3,
    74000: 4, 70000: 5,
                     70032: 6, 88000: 7, 86000: 8, 99000: 9,
    90000: 10, 93000: 11}

# Load test data
leaderboard_test_res = pd.read_hdf('ml_data.h5', '
    leaderboard_test_res')
hidden_test_res = pd.read_hdf('ml_data.h5', 'hidden_test_res')

leaderboard_test_target = pd.read_hdf('ml_data.h5', '
    leaderboard_test_features_target').values
leaderboard_test_target =  np.vectorize(lithology_numbers.get)(
    leaderboard_test_target)
hidden_test_target = pd.read_hdf('ml_data.h5', '
    hidden_test_target').values
hidden_test_target =  np.vectorize(lithology_numbers.get)(
    hidden_test_target)

leaderboard_accuracy_scores = []
hidden_accuracy_scores = []
for (leaderboard_column, leaderboard_data), (hidden_column,
    hidden_data) in zip(leaderboard_test_res.iteritems(),
    hidden_test_res.iteritems()):

    leaderboard_data =  np.vectorize(lithology_numbers.get)(
        leaderboard_data)
    leaderboard_accuracy_scores.append(np.around(score(
        leaderboard_data, leaderboard_test_target),4))
    hidden_data =  np.vectorize(lithology_numbers.get)(
        hidden_data)
```

```
33     hidden_accuracy_scores.append(np.around(score(hidden_data,
       hidden_test_target),4))
34
35 # plot the results
36 plt, ax1 = plt.subplots()
37 labels = leaderboard_test_res.columns
38 x = np.arange(len(labels))
39 width = 0.35
40 rects1 = ax1.bar(x - width/2, leaderboard_accuracy_scores, width,
       label='Leaderboard test data set', color='#C82127')
41 rects2 = ax1.bar(x + width/2, hidden_accuracy_scores, width,
       label='Hidden test est data set', color='#0A3A54')
42 ax1.set_ylabel('Accuracy scores')
43 ax1.set_ylim(0,-0.7)
44 ax1.set_xticks(x, labels)
45 ax1.legend()
46 ax1.bar_label(rects1, padding=-12)
47 ax1.bar_label(rects2, padding=-12)
```

Listing 8.15 Final scoring on the leaderbord and hidden test data set

References

Bestagini, P., Lipari, V., & Tubaro, S. (2017). A machine learning approach to facies classification using well logs. SEG Technical Program Expanded Abstracts, pp. 2137–2142. https://doi.org/10.1190/SEGAM2017-17729805.1

Bormann, P., Aursand, P., Dilib, F., Manral, S., & Dischington, P. (2020). FORCE 2020 Well well log and lithofacies dataset for machine learning competition. Dataset on Zenodo. https://doi.org/10.5281/ZENODO.4351156

Hall, B. (2016). Facies classification using machine learning. *Leading Edge, 35*(10), 906–909. https://doi.org/10.1190/TLE35100906.1/ASSET/IMAGES/LARGE/TLE35100906.1FIG2.JPEG

Hall, M., & Hall, B. (2017). Distributed collaborative prediction: Results of the machine learning contest. *The Leading Edge, 36*(3), 267–269. https://doi.org/10.1190/TLE36030267.1

Hernandez-Martinez, E., Perez-Muñoz, T., Velasco-Hernandez, J. X., Altamira-Areyan, A., & Velasquillo-Martinez, L. (2013). Facies recognition using multifractal hurst analysis: Applications to well-log data. *Mathematical Geosciences, 45*(4), 471–486. https://doi.org/10.1007/S11004-013-9445-6/FIGURES/9

Lemaître, G., Nogueira, F., & Aridas, C. K. (2017). Imbalanced-learn: A python toolbox to tackle the curse of imbalanced datasets in machine learning. *Journal of Machine Learning Research, 18*(17), 1–5.

Wood, D. A. (2021). Enhancing lithofacies machine learning predictions with gamma-ray attributes for boreholes with limited diversity of recorded well logs. *Artificial Intelligence in Geosciences, 2*, 148–164. https://doi.org/10.1016/J.AIIG.2022.02.007

Zou, Q., Ni, L., Zhang, T., & Wang, Q. (2015). Deep learning based feature selection for remote sensing scene classification. *IEEE Geoscience and Remote Sensing Letters, 12*(11), 2321–2325. https://doi.org/10.1109/LGRS.2015.2475299

Chapter 9
Machine Learning Regression in Petrology

9.1 Motivation

Deciphering magma storage depths and temperatures in the feeding systems of active volcanoes is a central issue in volcanology and petrology (see, e.g., Putirka, 2008). For example, magma storage depths help to characterize volcanic plumbing systems (see, e.g., Petrelli et al., 2018; Ubide and Kamber, 2018; Ubide et al., 2021). Also, the magma temperature must be estimated in order to use diffusion-based geo-chronometers (see, e.g., Costa et al., 2020). To date, a robust and widely applied strategy to design geo-barometers or geo-thermometers is mainly based on changes in entropy and volume during equilibrium reactions between melts and crystals (see Putirka, 2008 and Putirka, 2008, and references therein). For example, the calibration of a mineral-melt or mineral-only thermometer or of a barometer consists of five main steps: (1) determine the chemical equilibria associated with significant changes in entropy and volume (Putirka, 2008); (2) procure a suitable experimental data set for which temperature and pressure are known (e.g., the data set of the Library of Experimental Phase Relations; Hirschmann et al., 2008); (3) compute the components of the crystal phase from chemical analyses; (4) choose the regression strategy; and (5) validate the model (Putirka, 2008). Recently, numerous authors have demonstrated the potential of thermo-barometry based on ML (see, e.g., Petrelli et al., 2020; Jorgenson et al., 2022). This chapter discusses how to calibrate ML thermo-barometers based on ortopyroxenes in equilibrium with the melt phase and with orthopyroxenes alone.

M. Petrelli, *Machine Learning for Earth Sciences*, Springer Textbooks in Earth Sciences, Geography and Environment, https://doi.org/10.1007/978-3-031-35114-3_9

9.2   LEPR Data Set and Data Pre-processing

The Library of Experimental Phase Relations (LEPR) (Hirschmann et al., 2008) includes >5000 petrological experiments simulating igneous systems at temperatures between 500 and 2500 °C and pressures up to 25 GPa or more. The LEPR data set can be downloaded from a dedicated portal.[1] In the LEPR data set, the entries corresponding to each experiment include both experimental data (i.e., the composition of starting materials, the experimental temperature and pressure, the phases at the end of the experiments and related compositions) and metadata (e.g., author, laboratory, device, oxygen fugacity). For this chapter, I downloaded an Excel[TM] file and I named it LEPR_download.xls. In the Excel[TM] file, the sheet named "Experiments" contains all the meta data and relevant information such as the composition of starting materials, the experimental temperature and pressure, and the phases present at the end of the experiment. The sheets named with a phase name (e.g., Liquid, Clinopyroxene, Olivine) contain the chemical compositions for that specific phase in each experiment. An index characterizes each experiment, linking the information in the different sheets.

As a pre-processing strategy (see code listing 9.1), we define the function *data_pre_processing*(), which (1) imports the LEPR data set from Excel[TM] (lines 103 and 104), (2) creates a pandas *pipe*() for basic operations such as adjusting column names, converting all Fe data such as FeO_{tot}, filtering the features, and imputing NaN to zero (lines from 115 to 120); (3) start storing phase information in a .hd5 file (lines 123, 153, and 154); (4) combine all relevant data in a single pandas DataFrame (lines 128–130); (5) filter based on SiO_2, pressure P (GPa), and temperature T (°C) (lines 132–141); (6) remove the entries characterized by chemical analysis that do not fit the chemical formula of the orthopyroxene (lines 143–145); (7) shuffle the data set (lines 147 and 148); (8) separate the labels from the input features (lines 150 and 151); and (9) store everything in a .hd5 file (lines 153 and 154).

The statement at line 157 triggers the data pre-processing. The result is a hdf5 file named ml_data.h5 that contains a DataFrame named "Liquid_Orthopyroxene" hosting the pre-processed experimental data from the LEPR data set. In addition, it stores the labels T and P in a DataFrame named "labels." Finally, it contains all the original data of interest in three DataFrames named "Liquid," "Orthopyroxene," and "starting_material."

Figures 9.1 and 9.2 show the probability densities for the different chemical elements in the melt and orthopyroxene phases, respectively (code listing 9.2). Code listing 9.2 imports the Liquid_Orthopyroxene DataFrame from the hdf5 file ml_data.h5 (line 5).

---

[1] https://lepr.earthchem.org/.

```
 1  import os
 2  import pandas as pd
 3  import numpy as np
 4
 5  Elements  = {
 6    'Liquid': ['SiO2', 'TiO2', 'Al2O3', 'FeOtot', 'MgO',
 7               'MnO', 'CaO', 'Na2O', 'K2O'],
 8    'Orthopyroxene': ['SiO2', 'TiO2', 'Al2O3', 'FeOtot',
 9               'MgO', 'MnO', 'CaO', 'Na2O', 'Cr2O3'] }
10
11  def calculate_cations_on_oxygen_basis(
12          myData0, myphase, myElements, n_oxygens):
13
14      Weights = {'SiO2': [60.0843,1.0,2.0],
15                 'TiO2':[79.8788,1.0,2.0],
16                 'Al2O3': [101.961,2.0,3.0],
17                 'FeOtot':[71.8464,1.0,1.0],
18                 'MgO':[40.3044,1.0,1.0],
19                 'MnO':[70.9375,1.0,1.0],
20                 'CaO':[56.0774,1.0,1.0],
21                 'Na2O':[61.9789,2.0,1.0],
22                 'K2O':[94.196,2.0,1.0],
23                 'Cr2O3':[151.9982,2.0,3.0],
24                 'P2O5':[141.937,2.0,5.0],
25                 'H2O':[18.01388,2.0,1.0] }
26
27      myData = myData0.copy()
28      # Cation mole proportions
29      for el in myElements:
30          myData[el + '_cat_mol_prop'] = myData[myphase +
31                      '_' + el] * Weights[el][1] / Weights[el][0]
32      # Oxygen mole proportions
33      for el in myElements:
34          myData[el + '_oxy_mol_prop'] = myData[myphase +
35                      '_' + el] * Weights[el][2] / Weights[el][0]
36      # Oxigen mole proportions totals
37      totals = np.zeros(len(myData.index))
38      for el in myElements:
39          totals += myData[el + '_oxy_mol_prop']
40      myData['tot_oxy_prop'] = totals
41      # totcations
42      totals = np.zeros(len(myData.index))
43      for el in myElements:
44          myData[el + '_num_cat'] = n_oxygens * myData[el +
45                      '_cat_mol_prop']  /  myData['tot_oxy_prop']
46          totals += myData[el + '_num_cat']
47      return totals
48
49  def filter_by_cryst_formula(dataFrame, myphase, myElements):
50
51      c_o_Tolerance = {'Orthopyroxene': [4,6,0.025] }
52
53      dataFrame['Tot_cations'] = calculate_cations_on_oxygen_basis(
54          myData0 = dataFrame, myphase = myphase,
```

```
55            myElements = myElements,
56            n_oxygens = c_o_Tolerance[myphase][1])
57
58      dataFrame = dataFrame[
59          (dataFrame['Tot_cations'] < c_o_Tolerance[myphase][0]
60                              + c_o_Tolerance[myphase][2]) &
61          (dataFrame['Tot_cations'] > c_o_Tolerance[myphase][0]
62                              - c_o_Tolerance[myphase][2])]
63
64      dataFrame = dataFrame.drop(columns=['Tot_cations'])
65      return dataFrame
66
67  def adjustFeOtot(dataFrame):
68      for i in range(len(dataFrame.index)):
69          try:
70              if pd.to_numeric(dataFrame.Fe2O3[i])>0:
71                  dataFrame.loc[i,'FeOtot'] = (
72                      pd.to_numeric(dataFrame.FeO[i]) + 0.8998 *
73                      pd.to_numeric(dataFrame.Fe2O3[i]))
74              else:
75                  dataFrame.loc[i,
76                      'FeOtot'] = pd.to_numeric(dataFrame.FeO[i])
77          except:
78              dataFrame.loc[i,'FeOtot'] = 0
79      return dataFrame
80
81  def adjust_column_names(dataFrame):
82      dataFrame.columns = [c.replace('Wt: ', '')
83                          for c in dataFrame.columns]
84      dataFrame.columns = [c.replace(' ', '')
85                          for c in dataFrame.columns]
86      return dataFrame
87
88  def select_base_features(dataFrame, my_elements):
89      dataFrame = dataFrame[my_elements]
90      return dataFrame
91
92  def data_imputation(dataFrame):
93      dataFrame = dataFrame.fillna(0)
94      return dataFrame
95
96  def data_pre_processing(phase_1, phase_2, out_file):
97
98      try:
99          os.remove(out_file)
100     except OSError:
101         pass
102
103     starting = pd.read_excel('LEPR_download.xls',
104                          sheet_name='Experiment')
105     starting= adjust_column_names(starting)
106     starting.name = ''
107     starting = starting[['Index', 'T(C)','P(GPa)']]
108     starting.to_hdf(out_file, key='starting_material')
```

```
109
110    phases = [phase_1, phase_2]
111
112    for ix, my_phase in enumerate(phases):
113        my_dataset =  pd.read_excel('LEPR_download.xls',
114                                    sheet_name = my_phase)
115        my_dataset = (my_dataset.
116                        pipe(adjust_column_names).
117                        pipe(adjustFeOtot).
118                        pipe(select_base_features,
119                            my_elements= Elements[my_phase]).
120                        pipe(data_imputation))
121
122        my_dataset = my_dataset.add_prefix(my_phase + '_')
123        my_dataset.to_hdf(out_file, key=my_phase)
124
125    my_phase_1 = pd.read_hdf(out_file, phase_1)
126    my_phase_2 = pd.read_hdf(out_file, phase_2)
127
128    my_dataset = pd.concat([starting,
129                            my_phase_1,
130                            my_phase_2], axis=1)
131
132    my_dataset = my_dataset[(my_dataset['Liquid_SiO2'] > 35)&
133                            (my_dataset['Liquid_SiO2'] < 80)]
134
135    my_dataset = my_dataset[(
136        my_dataset['Orthopyroxene_SiO2'] > 0)]
137
138    my_dataset = my_dataset[(my_dataset['P(GPa)'] <= 2)]
139
140    my_dataset = my_dataset[(my_dataset['T(C)'] >= 650)&
141                            (my_dataset['T(C)'] <= 1800)]
142
143    my_dataset = filter_by_cryst_formula(dataFrame = my_dataset,
144                            myphase = phase_2,
145                            myElements = Elements[phase_2])
146
147    my_dataset = my_dataset.sample(frac=1,
148                        random_state=50).reset_index(drop=True)
149
150    my_labels = my_dataset[['Index', 'T(C)', 'P(GPa)']]
151    my_dataset = my_dataset.drop(columns=['T(C)','P(GPa)'])
152
153    my_labels.to_hdf(out_file, key='labels')
154    my_dataset.to_hdf(out_file,
155                        key= phase_1 + '_' + phase_2)
156
157 data_pre_processing(phase_1='Liquid' ,
158                     phase_2='Orthopyroxene',
159                     out_file='ml_data.h5')
```

Listing 9.1 Implementation of pre-processing strategy

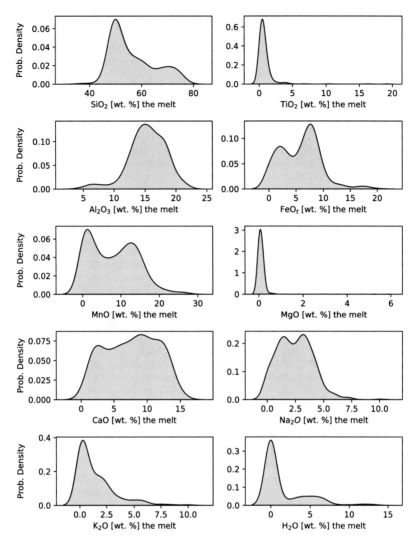

Fig. 9.1 Result of code listing 9.2. Descriptive statistics of the melt phase

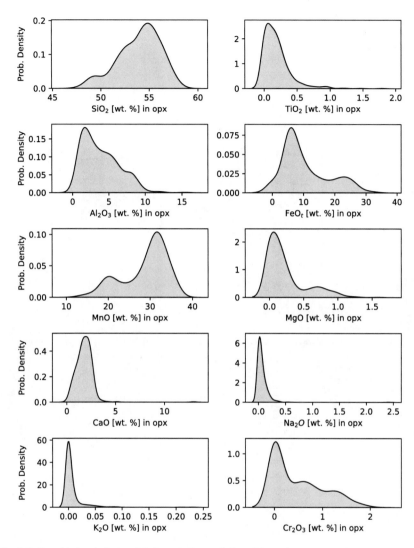

Fig. 9.2 Result of code listing 9.2. Descriptive statistics of the orthopyroxene phase

```
1  import pandas as pd
2  import matplotlib.pyplot as plt
3  import seaborn as sns
4
5  my_dataset = pd.read_hdf('ml_data.h5', 'Liquid_Orthopyroxene')
6
7  Elements = {
8    'Liquid': ['SiO2', 'TiO2', 'Al2O3', 'FeOtot', 'MgO',
9              'MnO', 'CaO', 'Na2O', 'K2O', 'H2O'],
10   'Orthopyroxene': ['SiO2', 'TiO2', 'Al2O3', 'FeOtot',
11              'MgO', 'MnO', 'CaO', 'Na2O', 'K2O', 'Cr2O3']}
12
13 fig = plt.figure(figsize=(7,9))
14 x_labels_melt = [r'SiO$_2$', r'TiO$_2$', r'Al$_2$O$_3$',
15                 r'FeO$_t$', r'MnO', r'MgO', r'CaO',
16                 r'Na$_2O$', r'K$_2$O', r'H$_2$O']
17 for i, col in enumerate(Elements['Liquid']):
18     ax1 = fig.add_subplot(5, 2, i+1)
19     sns.kdeplot(my_dataset['Liquid_' + col], fill=True,
20              color='k', facecolor='#BFD7EA', ax = ax1)
21     ax1.set_xlabel(x_labels_melt[i] + ' [wt. %] the melt')
22     if i in [0,2,4,6,8]:
23         ax1.set_ylabel('Prob. Density')
24     else:
25         ax1.set(ylabel=None)
26
27 fig.align_ylabels()
28 fig.tight_layout()
29
30 fig1 = plt.figure(figsize=(7,9))
31 x_labels_cpx = [r'SiO$_2$', r'TiO$_2$', r'Al$_2$O$_3$',
32                 r'FeO$_t$', r'MnO', r'MgO', r'CaO',
33                 r'Na$_2O$', r'K$_2$O', r'Cr$_2$O$_3$']
34 for i, col in enumerate(Elements['Orthopyroxene']):
35     ax2 = fig1.add_subplot(5, 2, i+1)
36     sns.kdeplot(my_dataset['Orthopyroxene_' + col], fill=True,
37              color='k', facecolor='#BFD7EA', ax = ax2)
38     ax2.set_xlabel(x_labels_cpx[i] + ' [wt. %] in opx')
39     if i in [0,2,4,6,8]:
40         ax2.set_ylabel('Prob. Density')
41     else:
42         ax2.set(ylabel=None)
43
44 fig1.align_ylabels()
45 fig1.tight_layout()
```

Listing 9.2 Descriptive statistics applied to orthopyroxenes

9.3 Compositional Data Analysis

In Sect. 3.3.6, we introduced the basic concept of compositional data analysis and discussed why most of the advanced statistical techniques cannot be applied to compositional data without a proper transformation. In fact, many statistical methods assume independent data in the range $-\infty$ to $+\infty$. Intrinsically, compositional features range from 0 to 100 (or from 0 to 1) and are not independent because changing the value of one element automatically affects the abundance of the other components (Aitchison, 1982). Decision-tree ensembles such as random forest (Song & Lu, 2015) and extremely randomized trees (Geurts et al., 2006) make no specific assumption about the data structure. Therefore, they can be applied to un-transformed data (Aitchison, 1982). However, recent studies report that tree ensembles perform better when applied to log-ratio pairwise-transformed data (Tolosana-Delgado et al., 2019). Although tree-based ensembles do not strictly require a CoDA transformation, they benefit from the introduction of new features (i.e., pairwise log-ratios) derived from existing features such as the augmentation of the feature input space. The result is reduced overfitting, which improves generalization. This chapter compares the results of the extremely randomized trees algorithm applied to both un-transformed and un-transformed plus log-ratio pairwise transformed data, as suggested by Tolosana-Delgado et al. (2019). To add the log-ratio pairwise transformation to our pre-processing strategy, we simply add a new function to code listing 9.1. Code listing 9.3 shows the final version of our pre-processing strategy, which now includes the log-ratio pairwise transformation.

```
1  import os
2  import pandas as pd
3  import numpy as np
4
5  Elements  = {
6    'Liquid': ['SiO2', 'TiO2', 'Al2O3', 'FeOtot', 'MgO',
7               'MnO', 'CaO', 'Na2O', 'K2O'],
8    'Orthopyroxene': ['SiO2', 'TiO2', 'Al2O3', 'FeOtot',
9               'MgO', 'MnO', 'CaO', 'Na2O', 'Cr2O3']}
10
11 def calculate_cations_on_oxygen_basis(
12        myData0, myphase, myElements, n_oxygens):
13
14    Weights = {'SiO2': [60.0843,1.0,2.0],
15               'TiO2':[79.8788,1.0,2.0],
16               'Al2O3': [101.961,2.0,3.0],
17               'FeOtot':[71.8464,1.0,1.0],
18               'MgO':[40.3044,1.0,1.0],
19               'MnO':[70.9375,1.0,1.0],
20               'CaO':[56.0774,1.0,1.0],
21               'Na2O':[61.9789,2.0,1.0],
22               'K2O':[94.196,2.0,1.0],
23               'Cr2O3':[151.9982,2.0,3.0],
```

```
24                    'P2O5':[141.937,2.0,5.0],
25                    'H2O':[18.01388,2.0,1.0]}
26
27       myData = myData0.copy()
28       # Cation mole proportions
29       for el in myElements:
30           myData[el + '_cat_mol_prop'] = myData[myphase +
31                       '_' + el] * Weights[el][1] / Weights[el][0]
32       # Oxygen mole proportions
33       for el in myElements:
34           myData[el + '_oxy_mol_prop'] = myData[myphase +
35                       '_' + el] * Weights[el][2] / Weights[el][0]
36       # Oxigen mole proportions totals
37       totals = np.zeros(len(myData.index))
38       for el in myElements:
39           totals += myData[el + '_oxy_mol_prop']
40       myData['tot_oxy_prop'] = totals
41       # totcations
42       totals = np.zeros(len(myData.index))
43       for el in myElements:
44           myData[el + '_num_cat'] = n_oxygens * myData[el +
45                       '_cat_mol_prop'] /  myData['tot_oxy_prop']
46           totals += myData[el + '_num_cat']
47       return totals
48
49  def filter_by_cryst_formula(dataFrame, myphase, myElements):
50
51       c_o_Tolerance = {'Orthopyroxene': [4,6,0.025]}
52
53       dataFrame['Tot_cations'] = calculate_cations_on_oxygen_basis(
54           myData0 = dataFrame, myphase = myphase,
55           myElements = myElements,
56           n_oxygens = c_o_Tolerance[myphase][1])
57
58       dataFrame = dataFrame[
59           (dataFrame['Tot_cations'] < c_o_Tolerance[myphase][0]
60                               + c_o_Tolerance[myphase][2]) &
61           (dataFrame['Tot_cations'] > c_o_Tolerance[myphase][0]
62                               - c_o_Tolerance[myphase][2])]
63
64       dataFrame = dataFrame.drop(columns=['Tot_cations'])
65       return dataFrame
66
67  def adjustFeOtot(dataFrame):
68       for i in range(len(dataFrame.index)):
69           try:
70               if pd.to_numeric(dataFrame.Fe2O3[i])>0:
71                   dataFrame.loc[i,'FeOtot'] = (
72                       pd.to_numeric(dataFrame.FeO[i]) + 0.8998 *
73                       pd.to_numeric(dataFrame.Fe2O3[i]))
74               else:
75                   dataFrame.loc[i,
76                       'FeOtot'] = pd.to_numeric(dataFrame.FeO[i])
77           except:
```

```
78                    dataFrame.loc[i,'FeOtot'] = 0
79        return dataFrame
80
81   def adjust_column_names(dataFrame):
82        dataFrame.columns = [c.replace('Wt: ', '')
83                              for c in dataFrame.columns]
84        dataFrame.columns = [c.replace(' ', '')
85                              for c in dataFrame.columns]
86        return dataFrame
87
88   def select_base_features(dataFrame, my_elements):
89        dataFrame = dataFrame[my_elements]
90        return dataFrame
91
92   def data_imputation(dataFrame):
93        dataFrame = dataFrame.fillna(0)
94        return dataFrame
95
96   def pwlr(dataFrame, my_phases):
97
98        for my_pahase in my_phases:
99            my_indexes  = []
100           column_list = Elements[my_pahase]
101
102           for col in column_list:
103               col = my_pahase + '_' + col
104               my_indexes.append(dataFrame.columns.get_loc(col))
105               my_min = dataFrame[col][dataFrame[col] > 0].min()
106               dataFrame.loc[dataFrame[col] == 0,
107                   col] = dataFrame[col].apply(
108                   lambda x: np.random.uniform(
109                       np.nextafter(0.0, 1.0),my_min))
110
111           for ix in range(len(column_list)):
112               for jx in range(ix+1, len(column_list)):
113                   col_name = 'log_' + dataFrame.columns[
114                       my_indexes[jx]] + '_' + dataFrame.columns[
115                           my_indexes[ix]]
116                   dataFrame.loc[:,col_name] =  np.log(
117                       dataFrame[dataFrame.columns[my_indexes[jx]]]/ \
118                       dataFrame[dataFrame.columns[my_indexes[ix]]])
119       return dataFrame
120
121  def data_pre_processing(phase_1, phase_2, out_file):
122
123       try:
124           os.remove(out_file)
125       except OSError:
126           pass
127
128       starting = pd.read_excel('LEPR_download.xls',
129                                sheet_name='Experiment')
130       starting= adjust_column_names(starting)
131       starting.name = ''
```

```
132    starting = starting[['Index', 'T(C)','P(GPa)']]
133    starting.to_hdf(out_file, key='starting_material')
134
135    phases = [phase_1, phase_2]
136
137    for ix, my_phase in enumerate(phases):
138        my_dataset = pd.read_excel('LEPR_download.xls',
139                                   sheet_name = my_phase)
140
141        my_dataset = (my_dataset.
142                          pipe(adjust_column_names).
143                          pipe(adjustFeOtot).
144                          pipe(select_base_features,
145                              my_elements= Elements[my_phase]).
146                          pipe(data_imputation))
147
148        my_dataset = my_dataset.add_prefix(my_phase + '_')
149        my_dataset.to_hdf(out_file, key=my_phase)
150
151    my_phase_1 = pd.read_hdf(out_file, phase_1)
152    my_phase_2 = pd.read_hdf(out_file, phase_2)
153
154    my_dataset = pd.concat([starting,
155                            my_phase_1,
156                            my_phase_2], axis=1)
157
158    my_dataset = my_dataset[(my_dataset['Liquid_SiO2'] > 35)&
159                            (my_dataset['Liquid_SiO2'] < 80)]
160
161    my_dataset = my_dataset[(
162        my_dataset['Orthopyroxene_SiO2'] > 0)]
163
164    my_dataset = my_dataset[(my_dataset['P(GPa)'] <= 2)]
165
166    my_dataset = my_dataset[(my_dataset['T(C)'] >= 650)&
167                            (my_dataset['T(C)'] <= 1800)]
168
169    my_dataset = filter_by_cryst_formula(dataFrame = my_dataset,
170                                   myphase = phase_2,
171                                   myElements = Elements[phase_2])
172
173    my_dataset = my_dataset.sample(frac=1,
174                            random_state=50).reset_index(drop=True)
175
176    my_labels = my_dataset[['Index', 'T(C)', 'P(GPa)']]
177    my_dataset = my_dataset.drop(columns=['T(C)','P(GPa)'])
178
179    my_labels.to_hdf(out_file, key='labels')
180    my_dataset.to_hdf(out_file, key= phase_1 + '_' + phase_2)
181
182    my_dataset = pwlr(my_dataset,
183                            my_phases= [phase_1, phase_2])
184    my_dataset.to_hdf(out_file,
185                      key= phase_1 + '_' + phase_2 + '_lrpwt')
```

```
186
187 data_pre_processing(phase_1='Liquid' ,
188                     phase_2='Orthopyroxene',
189                     out_file='ml_data.h5')
```

Listing 9.3 Final implementation of our pre-processing strategy

9.4 Model Training and Error Assessment

In agreement with Petrelli et al. (2020), we train the extremely randomized trees algorithm on the pre-processed data set. Also, we use a Monte Carlo simulation to propagate the errors and assess the goodness of the model. The Monte Carlo approach consists of repeating many times (i) the random splitting of the data set, and (ii) the training of the algorithm starting from a different random seeding (code listing 9.4). To achieve our goal, we define a function named *monte_carlo_simulation*() (line 9). Within this function, we repeat the train-validation splitting *n* times (lines 16–18), normalization to zero mean and unit variance (lines 20–22), training (lines 24–26), prediction (line 27), error assessment (lines 29–35), and the storing of the results (lines 36–42).

```
1 import pandas as pd
2 import numpy as np
3 from sklearn.preprocessing import StandardScaler
4 from sklearn.ensemble import ExtraTreesRegressor
5 from sklearn.model_selection import train_test_split
6 from sklearn.metrics import r2_score
7 from sklearn.metrics import mean_squared_error
8
9 def monte_carlo_simulation(X, y, indexes, n, key_res):
10
11     r2 = []
12     RMSE = []
13
14     for i in range(n):
15         my_res = {}
16         X_train, X_valid, y_train, y_valid, \
17             indexes_train, indexes_valid = train_test_split(
18                 X, y.ravel(), indexes, test_size=0.2)
19
20         scaler = StandardScaler().fit(X_train)
21         X_train = scaler.transform(X_train)
22         X_valid = scaler.transform(X_valid)
23
24         regressor = ExtraTreesRegressor(n_estimators=450,
25                                         max_features=1).fit(
26                                             X_train, y_train)
```

```
27      my_prediction = regressor.predict(X_valid)
28
29      my_res = {'indexes_valid': indexes_valid,
30                'prediction': my_prediction}
31
32      my_res_pd = pd.DataFrame.from_dict(my_res)
33      r2.append(r2_score(y_valid, my_prediction))
34      RMSE.append(np.sqrt(mean_squared_error(y_valid,
35                                             my_prediction)))
36      my_res_pd.to_hdf('ml_data.h5',
37                       key= key_res + '_res_' + str(i))
38
39   my_scores = {'r2_score': r2,
40               'root_mean_squared_error': RMSE}
41   my_scores_pd = pd.DataFrame.from_dict(my_scores)
42   my_scores_pd.to_hdf('ml_data.h5', key = key_res + '_scores')
43
44
45 my_keys = ['Liquid_Orthopyroxene', 'Liquid_Orthopyroxene_1rpwt']
46
47 for my_key in my_keys:
48
49    # Liquid plus opx calibration
50    liquid_opx = pd.read_hdf('ml_data.h5', my_key)
51    print(liquid_opx.columns)
52    X_liquid_opx = liquid_opx.values
53    my_labels = pd.read_hdf('ml_data.h5', 'labels')
54    my_y = my_labels['T(C)'].values
55    my_indexes = my_labels['Index'].values
56    monte_carlo_simulation(X = X_liquid_opx, y = my_y,
57                           indexes = my_indexes,
58                           n =1000, key_res = my_key)
59
60    # opx only calibration
61    opx = liquid_opx.loc[:,
62               ~liquid_opx.columns.str.startswith('Liquid')]
63    X_opx = opx.values
64    my_key = my_key.replace("Liquid_", "")
65    monte_carlo_simulation(X = X_opx,
66                           y = my_y, indexes = my_indexes,
67                           n =1000, key_res = my_key)
```

Listing 9.4 Training of the model in a Monte Carlo simulation

9.5   Evaluation of Results

Figures 9.3 and 9.4 show the results of the Monte Carlo simulations (derived from code listing 9.5); the upper panels refer to raw data, whereas the lower panels

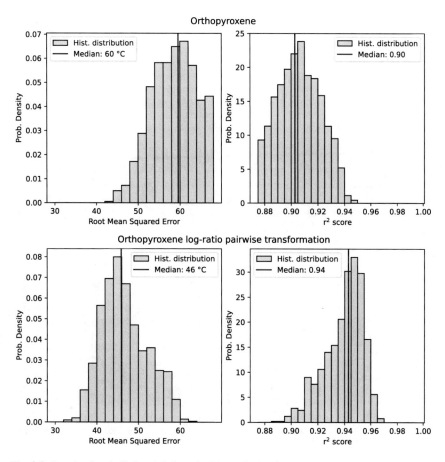

Fig. 9.3 Result of code listing 9.5 (i.e., the Monte Carlo simulation of the orthopyroxene-only system)

display the results on raw data plus the features deriving from the log-ratio pairwise transformation.

Note that adding the features deriving from the log-ratio pairwise transformation seems improving the performance of the orthopyroxene-only calibration of the thermometer (Fig. 9.3). In this case, the root mean squared error and r^2 improve by 14 °C and 0.4, respectively.

In contrast with the orthopyroxene-only calibration, the liquid plus orthopyroxene system does not benefit from the addition of features deriving from the log-ratio pairwise transformation (Fig. 9.4). In this case, the root mean squared error only differs by 4 °C and r^2 is stable at 0.95.

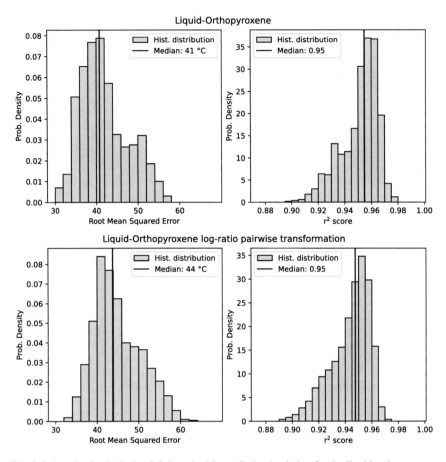

Fig. 9.4 Result of code listing 9.5 (i.e., the Monte Carlo simulation for the liquid-orthopyroxene system)

```
import pandas as pd
import numpy as np
import matplotlib.pyplot as plt

for my_key in ['Orthopyroxene', 'Liquid_Orthopyroxene']:

    fig = plt.figure(figsize=(8,8),constrained_layout=True)
    subfigs = fig.subfigures(nrows=2, ncols=1)
    for j, (trans, my_title) in enumerate(zip(['', '_lrpwt'],
        [my_key, my_key+' log-ratio pairwise transformation'])):
        my_scores = pd.read_hdf('ml_data.h5',
                                my_key + trans + '_scores')

        RMSE_ML_valid_median_T = np.median(
            my_scores['root_mean_squared_error'])
        R2_valid_median_T = np.median(my_scores['r2_score'])
```

```
17
18      subfigs[j].suptitle(my_title.replace('_', '-'))
19
20      # left panel
21      ax = subfigs[j].add_subplot(1, 2,1)
22      bins = np.arange(30, 70, 2)
23      ax.hist(my_scores['root_mean_squared_error'], bins=bins,
24              density = True, color = '#BFD7EA',
25              edgecolor = 'k',
26              label='Hist. distribution')
27      ax.axvline(RMSE_ML_valid_median_T,
28              color='#C82127',
29              label='Median: {:.0f}  C'.format(
30                  RMSE_ML_valid_median_T))
31      ax.set_xlabel('Root Mean Squared Error')
32      ax.set_ylabel('Prob. Density')
33      ax.legend()
34
35      # right panel
36      ax = subfigs[j].add_subplot(1, 2, 2)
37      bins = np.arange(0.875, 1, 0.005)
38      ax.hist(my_scores['r2_score'], bins = bins,
39              density = True, color = '#BFD7EA',
40              edgecolor='k',
41              label='Hist. distribution')
42      ax.axvline(R2_valid_median_T, color='#C82127',
43              label='Median: {:.2f}'.format(
44                  R2_valid_median_T))
45      ax.set_xlabel(r'r$^2$ score')
46      ax.set_ylabel('Prob. Density')
47      ax.legend()
```

Listing 9.5 Plots the results of the Monte Carlo simulation

References

Aitchison, J. (1982). The statistical analysis of compositional data. *Journal of the Royal Statistical Society. Series B (Methodological), 44*(2), 139–177.

Costa, F., Shea, T., & Ubide, T. (2020). Diffusion chronometry and the timescales of magmatic processes. *Nature Reviews Earth and Environment, 1*(4), 201–214. https://doi.org/10.1038/s43017-020-0038-x

Geurts, P., Ernst, D., & Wehenkel, L. (2006). Extremely randomized trees. *Machine Learning, 63*(1), 3–42. https://doi.org/10.1007/S10994-006-6226-1

Hirschmann, M., Ghiorso, M., Davis, F., Gordon, S., Mukherjee, S., Grove, T., Krawczynski, M., Medard, E., & Till, C. (2008). Library of experimental phase relations (LEPR): A database andWeb portal for experimental magmatic phase equilibria data. *Geochemistry, Geophysics, Geosystems, 9*(3). https://doi.org/10.1029/2007GC001894

Jorgenson, C., Higgins, O., Petrelli, M., Bégué, F., & Caricchi, L. (2022). A machine learning-based approach to clinopyroxene thermobarometry: Model optimization and distribution for use in earth sciences. *Journal of Geophysical Research: Solid Earth, 127*(4), e2021JB022904. https://doi.org/10.1029/2021JB022904

Petrelli, M., Caricchi, L., & Perugini, D. (2020). Machine learning thermo-barometry: Application to clinopyroxene-bearing magmas. *Journal of Geophysical Research: Solid Earth, 125*(9). https://doi.org/10.1029/2020JB020130

Petrelli, M., El Omari, K., Spina, L., Le Guer, Y., La Spina, G., & Perugini, D. (2018). Timescales of water accumulation in magmas and implications for short warning times of explosive eruptions. *Nature Communications, 9*(1), 770. https://doi.org/10.1038/s41467-018-02987-6

Putirka, K. (2008). Thermometers and barometers for volcanic systems. *Reviews in Mineralogy and Geochemistry, 69*(1), 61–120. https://doi.org/10.2138/rmg.2008.69.3

Song, Y. Y., & Lu, Y. (2015). Decision tree methods: Applications for classification and prediction. *Shanghai Archives of Psychiatry, 27*(2), 130. https://doi.org/10.11919/J.ISSN.1002-0829.215044

Tolosana-Delgado, R., Talebi, H., Khodadadzadeh, M., & Boogaart, K. G. (2019). On machine learning algorithms and compositional data. In Proceedings of the 8th International Workshop on Compositional Data Analysis (CoDaWork2019) (pp. 172–175).

Ubide, T., & Kamber, B. (2018). Volcanic crystals as time capsules of eruption history. *Nature Communications, 9*(1), 326. https://doi.org/10.1038/s41467-017-02274-w

Ubide, T., Neave, D., Petrelli, M., & Longpré, M.-A. (2021). Editorial: Crystal archives of magmatic processes. *Frontiers in Earth Science, 9*. https://doi.org/10.3389/feart.2021.749100

Part IV
Scaling Machine Learning Models

Chapter 10
Parallel Computing and Scaling with Dask

10.1 Warming Up: Basic Definitions

Processor, CPU, Core The traditional definition of "processor" and "central processing unit" (CPU) is "a microprocessor chip that sequentially (i.e., one by one) executes a series of basic processing tasks based on an input" (Caesar Wu, 2015). However, modern CPUs largely exceed this traditional definition by integrating many components and hosting a cache memory. Modern CPUs duplicate and execute the most basic processing tasks by applying self-contained execution blocks that fit the traditional definition of a processor (Caesar Wu, 2015). These self-contained execution blocks are typically called "cores" (Caesar Wu, 2015).

Multi-Core Processor and Parallel Hardware Multi-core processors, chip multi-processors (CMPs), and parallel hardware are often used as synonyms (Peter Pacheco, 2020). A CMP incorporates many processors and cache memory on a chip. Parallel hardware is ubiquitous now—it is almost impossible to find a modern laptop, desktop, or server that does not use a multi-core processor (Peter Pacheco, 2020).

Graphics Processing Unit (GPU) "GPUs are multi-core processing units made of massively parallel, smaller, and more specialized cores than those generally found in high-performance CPUs. GPU architecture efficiently processes vector data (an array of numbers) and is often referred to as vector architecture."[1]

Field Programmable Gate Array (FPGA) "FPGAs are integrated circuits with a programmable hardware fabric. Unlike CPUs and GPUs, which are software-

---

[1] https://intel.ly/39XimzH.

M. Petrelli, *Machine Learning for Earth Sciences*, Springer Textbooks
in Earth Sciences, Geography and Environment,
https://doi.org/10.1007/978-3-031-35114-3_10

programmable fixed architectures, FPGAs are reconfigurable. When writing software for a FPGA, compiled instructions become hardware components that are spatially laid out on the FPGA fabric, and those components can all execute in parallel." (see footnote 1).

Distributed Computing Distributed computing is "[a] computer system consisting of a multiplicity of processors, each with its own local memory, connected via a network. Loading or store instructions issued by a processor can only address the local memory and different mechanisms are provided for global communication" (David, 2011).

Serial Codes Serial codes are codes that were conceived and written for a single processor (Peter Pacheco, 2020). If you run a serial code on multiple processors or a distributed architecture, the performance does not magically improve because the instructions are executed sequentially by one of the available cores.

Parallel Computing Parallel computing is a computation strategy whereby many calculations or processes are executed simultaneously (Peter Pacheco, 2020). Parallel computing exploits multiple processors (i.e., CMP, GPU, and FPGA) or a distributed architecture (Peter Pacheco, 2020).

10.2   Basics of Dask

The goal of Dask[2] is to overcome single-machine restrictions by adding object scalability to Python scientific libraries such as pandas, NumPy, and scikit-learn (Daniel, 2019). Dask consists of three main layers: (1) scheduler, (2) low-level application programming interfaces (APIs), and (3) high-level APIs (Fig. 10.1). This chapter discusses mainly the high-level APIs that govern Dask arrays, Dask DataFrames, and Dask ML, which allow us to scale NumPy, pandas, and scikit-learn objects, respectively. In using Dask, our main goal is to extend the capabilities of single machines so that they can work with data sets that exceed their native RAM capabilities and deploy clusters to exploit big data sets or extremely complex models.

Dask Array
Dask arrays combine many NumPy arrays arranged into chunks (i.e., a single NumPy array) within a grid (Fig. 10.2). They are the parallel-friendly version of NumPy arrays. Defining a Dask array is as simple as defining a NumPy array; the only difference being that you need to import dask.array instead of NumPy (Fig. 10.3). For example, Fig. 10.3 shows how to create a $10^5 \times 10^5$ Dask array containing random numbers. In Jupyter Notebooks, you can easily retrieve copious information on the Dask array you created (Fig. 10.3).

---

[2] https://www.dask.org.

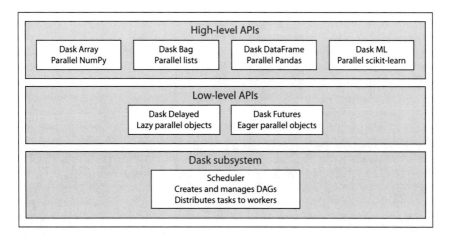

Fig. 10.1 Dask fundamentals, modified from (Daniel, 2019)

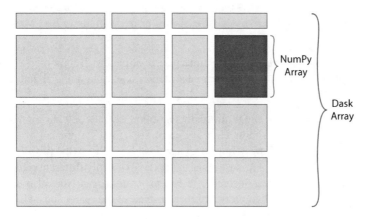

Fig. 10.2 Dask arrays, modified from https://examples.dask.org/array.html

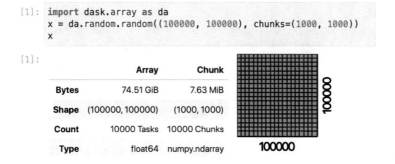

Fig. 10.3 Defining a Dask array

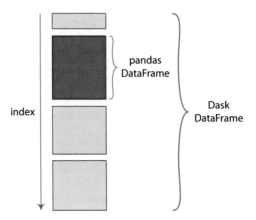

Fig. 10.4 Dask arrays, modified from https://examples.dask.org/dataframe.html

For example, Fig. 10.3 shows that the total size of x is 74.51 GiB (i.e., Gibibytes, GiB, with 1 GiB $\approx$ 1.074 GB). Also, the size of a single chunk is 7.63 MiB.

Dask Data Frame

A Dask DataFrame is the parallel counterpart of a pandas Dataframe (Fig. 10.4). They are composed of many smaller pandas DataFrames split along an index (Table 10.1).

To see how to use Dask DataFrames, let us import the data set that we developed in Chap. 8 and that we saved as HDF5. Figure 10.5 shows a portion of a Jupyter Notebook and highlights how to import a Dask DataFrame from the file *ml_data.h5*. Note that the procedure is similar to that in pandas. The only difference consists of importing a *dask.DataFrame* instead of a *pandas.DataFrame*. Note also that Dask splits the DataFrame into two parts and that, instead of the real values, all rows are filled with ellipses (...). This is because the data set is subject to "lazy" evaluation (see Sect. 10.3 for further details). To physically import *train_dataset*, Dask requires the additional step of using the *compute()* method (Fig. 10.6).

Dask ML

Model scaling can solve two common issues related to (1) data size and (2) model size (Table 10.2, Fig. 10.7). Pandas, NumPy, and scikit-learn are the libraries of choice to develop a ML strategy when your data set comfortably fits the free RAM of your computing environment (i.e., you are working with a small data set; see Table 10.2). In this case, scaling along the x dimension of Fig. 10.7 is not required and not recommended.

As an example, code listing in Fig. 10.8 shows how to use Numpy to define (line 2) a small data set *my_data* composed of 10^8 normally distributed pseudo-random numbers characterized by a mean value and standard deviation of one and two, respectively. Lines 3 and 4 simply check that the mean and the standard deviation

Table 10.1 Dask methods to import and create a Dask DataFrame. Please note that most of them are equivalent to pandas methods, i.e., Table 3.1 (modified from https://docs.dask.org/en/stable/dataframe-api.html)

Method	Description
read_table()	Read general delimited file
read_csv()	Read comma-separated values (csv) files
read_fwf()	Read fixed-width files
read_parquet()	Read parquet files
read_hdf()	Read Hierarchical Data Format (HDF) files
read_json()	Create a Dask DataFrame from a set of JSON files
read_orc()	Create a Dask DataFrame from ORC file(s)
read_sql_table()	Read SQL database table
read_sql_query()	Read SQL query
read_sql()	Read SQL query or database table
from_array()	Read any sliceable array
from_bcolz()	Read BColz CTable
from_dask_array()	Create a Dask DataFrame from a Dask Array
from_delayed()	Create a Dask DataFrame from many Dask Delayed objects
from_map()	Create a Dask DataFrame collection from a custom function map
from_pandas()	Construct a Dask DataFrame from a Pandas DataFrame
from_dict()	Construct a Dask DataFrame from a Python Dictionary
Bag_to_dataframe()	Create Dask Dataframe from a Dask Bag

```
[1]: import dask.dataframe as dd

[3]: train_dataset = dd.read_hdf('ml_data.h5', key='train')

[4]: train_dataset
```

[4]: **Dask DataFrame Structure:**

	CALI	Delta_CALI	log_RMED	Delta_log_RMED	log_RDEP	Delta_log_RDEP
npartitions=2						
	float64	float64	float64	float64	float64	float64
	...	...	...	...	...	...
	...	...	...	...	...	...

Dask Name: read-hdf, 2 tasks

Fig. 10.5 Importing a pandas DataFrame stored in an HDF5 files as a Dask DataFrame

of *my_data* are one and two, respectively. Finally, line 5 estimated the memory required by *my_data*, which is approximately 0.745 GiB.

However, when the size of the data set reaches the upper bound of the RAM (including any virtual memory generated by using the hard disk), memory errors start occurring (see code listing in Fig. 10.9). For example, increasing the size of *my_data* to 2.5×10^9 in a Linux system with 16 GB of free memory produces a "Memory error" because the operating system is "Unable to allocate 18.6 GiB for

```
[5]: train_dataset.compute()
```

[5]:		CALI	Delta_CALI	log_RMED	Delta_log_RMED	log_RDEP	Delta_log_RDEP
	0	19.480835	0.000000	0.207206	0.000000	0.254954	0.000000
	1	19.468800	-0.012035	0.208997	0.001791	0.254220	-0.000735
	2	19.468800	0.000000	0.211243	0.002246	0.255449	0.001230
	3	19.459282	-0.009518	0.209942	-0.001301	0.255638	0.000189
	4	19.453100	-0.006182	0.204847	-0.005096	0.254137	-0.001501
	...	...	...	...	...	...	...
	1170506	8.423170	0.001802	0.247442	0.000026	0.241466	0.000024
	1170507	8.379244	-0.043926	0.247442	0.000026	0.241466	0.000024
	1170508	8.350248	-0.028996	0.247442	0.000026	0.241466	0.000024
	1170509	8.313779	-0.036469	0.247442	0.000026	0.241466	0.000024
	1170510	8.294910	-0.018868	0.247442	0.000026	0.241466	0.000024

1170511 rows × 26 columns

Fig. 10.6 Physically importing a pandas DataFrame stored in an HDF5 files as Dask DataFrame

Table 10.2 Data set classification as a function of data size. Modified from Daniel (2019)

Data set size	Approximate size range	Fits in RAM?	Fits on local disk?
Small data set	Less than the free RAM on your system (e.g., 16 GB)	Yes	Yes
Medium data set	Larger than the free RAM on your system and less than capacity of the local disk (e.g., 2 TB)	No	Yes
Large data set	Larger than the capacity of the local disk	No	No

an array with shape (2 500 000 000) and data type float64'. This is clearly a data size issue because I generated a "medium data set" (see Table 10.2).

The use of Dask arrays allows you to overcome the problem with minimal changes in the code. For example, code listing in Fig. 10.10 uses Dask arrays (i.e., the parallel mimic of NumPy arrays) on a Lunix OS with 16 GB of free ram to complete the simple operations that were previously impossible using NumPy (i.e., code listing 10.9).

When model size is the problem (e.g, the model is growing too much or becoming too complex), all computations take extremely long. For example, the grid search done in Chap. 8 took several hours to complete. While waiting a few hours may not a be a big problem, the execution time will drastically increase up to days or even weeks upon simply increasing the dimension of the grid search (e.g.,

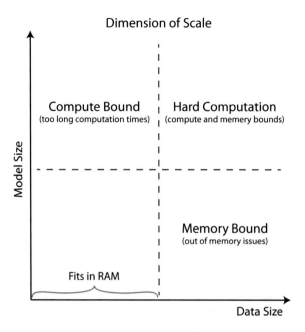

Fig. 10.7 Dimension of scale, modified from https://ml.dask.org

```
In [1]: import numpy as np

In [2]: my_dataset = np.random.normal(loc=1.0, scale=2.0, size=100000000)

In [3]: np.mean(my_dataset)
Out[3]: 0.9995566509046069

In [4]: np.std(my_dataset)
Out[4]: 1.9999512502789483

In [5]: my_dataset.nbytes / 1024**3
Out[5]: 0.7450580596923828
```

Fig. 10.8 Working with a small data set (i.e., well-fitting your RAM budget)

increasing the number of investigated hyper-parameters and densifying the grid) or the complexity of the decision tree ensemble (e.g., increasing the number of estimators). To optimize several ML models, the total time required can easily be on the order of months or even years.

The main aim of Dask ML is thus to provide scalable ML in Python for popular ML libraries such as scikit-Learn (Pedregosa et al., 2011), XGBoost, and others.

```
In [1]:  import numpy as np

In [2]:  my_dataset = np.random.normal(loc=1.0, scale=2.0, size=2500000000)

         ---------------------------------------------------------------
         -------
         MemoryError                              Traceback (most recent call
         last)
         Input In [2], in <module>
         ----> 1 my_dataset = np.random.normal(loc=1.0, scale=2.0, size=250000
         0000)

         File mtrand.pyx:1507, in numpy.random.mtrand.RandomState.normal()

         File _common.pyx:598, in numpy.random._common.cont()

         MemoryError: Unable to allocate 18.6 GiB for an array with shape (250
         0000000,) and data type float64
```

Fig. 10.9 When you exceed the free memory, you get a "Memory error"

```
In [2]:  import dask.array as da

In [4]:  my_dataset = da.random.normal(loc=1.0, scale=2.0, size=2500000000)

In [7]:  da.mean(my_dataset).compute()
Out[7]:  1.0000155698953217

In [8]:  da.std(my_dataset).compute()
Out[8]:  2.000041110411836
```

Fig. 10.10 Using Dask to work with a medium size data set

10.3   Eager Computation Versus Lazy Evaluation

Python usually uses the so-called "eager" computation, which simply means that Python immediately performs each operation such as transformations and calculations. For example, Fig. 10.11 shows the definition of the eager function *simple_lithopress*() (line 2) that estimates the lithostatic pressure assuming both the density and acceleration due to gravity are constants. We disclose the eager nature of the function at lines 3 and 4, since *simple_lithopress*() returns a computed value as soon as we call it in the code workflow; in other words, the calculations is done immediately.

```
[1]:  def simple_lithopress(z, ro, g):
          p_MPa = z*g*ro/1e6 # return the pressure in MPa
          return p_MPa

[3]:  my_pressure = simple_lithopress(z=2000, ro=2900, g=9.8)

[3]:  my_pressure

[3]:  56.84
```

Fig. 10.11 Defining the eager function *simple_lithopress*()

```
[4]: import numpy as np

[5]: %%time
     z_dist = np.random.normal(loc=2000, scale=200, size= 10000000)
     ro_dist = np.random.normal(loc=2900, scale=290, size= 10000000)
     g_dist = np.random.normal(loc=9.8, scale=0.1, size= 10000000)
     my_pressure_dist = simple_lithopress(z=z_dist, ro = ro_dist, g = g_dist)

     CPU times: user 854 ms, sys: 106 ms, total: 960 ms
     Wall time: 963 ms

[6]: my_pressure_dist.nbytes / 1024**2 # size in MB

[6]: 76.2939453125
```

Fig. 10.12 Performing a Monte Carlo error propagation using the 'Eager' *simple_lithopress*()

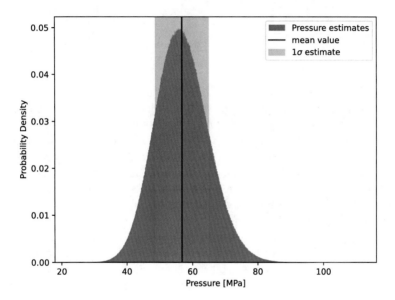

Fig. 10.13 Result of code listing 10.1

Similarly, if we perform a Monte Carlo error propagation (Fig. 10.12) combining NumPy arrays and the *simple_lithopress*() function, we get an immediate execution lasting less than one second and that generates an array of 10^7 elements ($\approx$76 MB).

To be aware of what we are doing, Fig. 10.13 shows the distribution of the computed pressures resulting from estimates of depth, density, and acceleration due to gravity and also accounting for the error estimates.

```
 1 import matplotlib.pyplot as plt
 2
 3 my_pressure_mean = np.mean(my_pressure_dist)
 4 my_pressure_std = np.std(my_pressure_dist)
 5
 6 fig, ax = plt.subplots()
 7 ax.hist(my_pressure_dist, density=True, bins='auto',
 8         color='#0F7F8B', label='Pressure estimates')
 9 ax.axvline(my_pressure_mean, color='#C82127', label='mean value')
10 ax.axvspan(my_pressure_mean - my_pressure_std,
11            my_pressure_mean + my_pressure_std,
12            color='#F15C61', alpha=0.4,
13            label=r'1$\sigma$ estimate')
14 ax.set_xlabel('Pressure [MPa]')
15 ax.set_ylabel('Probability Density')
16 ax.legend()
17 plt.show()
```

Listing 10.1 Plotting the results of the Monte Carlo error propagation.

Lazy evaluation differs from eager computation. Under lazy evaluation, Dask prepares a directed acyclic graph (DAG) for the functions, operations, and transformations involved. But it does not perform any computation. DAGs are mathematical objects deriving from graph theory. The theory behind DAGs and graph theory is outside the scope of this book, so please refer to specialized references to go learn the details of DAGs (Xu, 2003; Fiore & Campos, 2013; Maurer, 2013).

This section focuses mainly on learning the main benefits of using DAGs for our computations. One of the most important benefits is that the structure and the complexity of your computations can be evaluated and visualized before running them, which brings many advantages. For example, it allows you to decide whether to run your code on a single machine, a small cluster, or a high-performance computing facility. Figure 10.14 shows how to perform a lazy evaluation of the

```
[1]: import dask.array as da
     from dask import delayed

[2]: def simple_lithopress(z, ro, g):
         p_MPa = z*g*ro/1e6 # return the pressure in MPa
         return p_MPa

     z_da = da.random.normal(loc=2000, scale=200, size= 10000000)
     ro_da = da.random.normal(loc=2900, scale=290, size= 10000000)
     g_da = da.random.normal(loc=9.8, scale=0.1, size= 10000000)

     my_pressure_da = da.map_blocks(simple_lithopress, z_da, g_da, ro_da)

[3]: my_pressure_da = simple_lithopress(z=z_da, ro = ro_da, g = g_da)

[4]: my_pressure_da.visualize(filename='my_DAG.pdf')
```

Fig. 10.14 How to visualize a DAG in Dask

Fig. 10.15 A simple DAG
resulting from the code listing
reported above

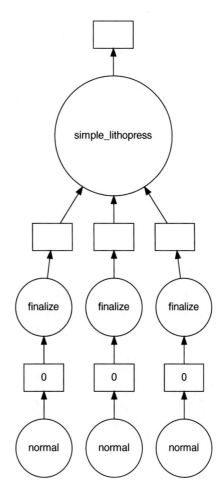

Monte Carlo error propagation performed in Fig. 10.12, and the resulting DAG is
shown in Fig. 10.15. It is a simple structure showing that, after generating three
normal distributions for the depth, density, and acceleration due to gravity, the
simple_litohpress() function uses them as input and generates an output. If we
increase the size of the three input arrays from 10^7 to 10^8, the structure of the
DAG changes (Fig. 10.16). In detail, we defined a so-called "embarassingly parallel"
workload (Fig. 10.17).

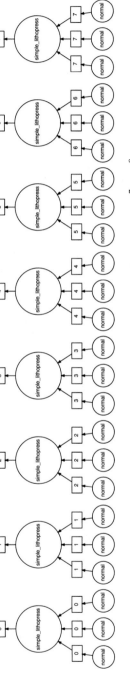

Fig. 10.16 The DAG that I obtain in my MacBook Pro if I increase the dimension of the input arrays from 10^7 to 10^8

Fig. 10.17 "Embarassingly
parallel" workload

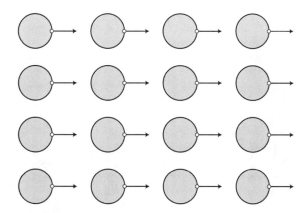

10.4 Diagnostic and Feedback

The Dask distributed scheduler provides an effective interactive dashboard that
consists of a rich ecosystem of monitoring and profiling tools that can be accessed by
a web browser (Fig. 10.18). The left and right panels of Fig. 10.18 display a Jupyter
Notebook and the Dask interactive dashboard, respectively. The Jupyter Notebook
starts the Dask client and its interactive dashboard at line 2 and then defines (lines
3 and 4), evaluates (line 5), and finally triggers (line 6, in progress and therefore
displayed as *) the computations. The right portion of the monitor shows the Dask
interactive dashboard during the ongoing process triggered by the Jupyter Notebook
at line 6.

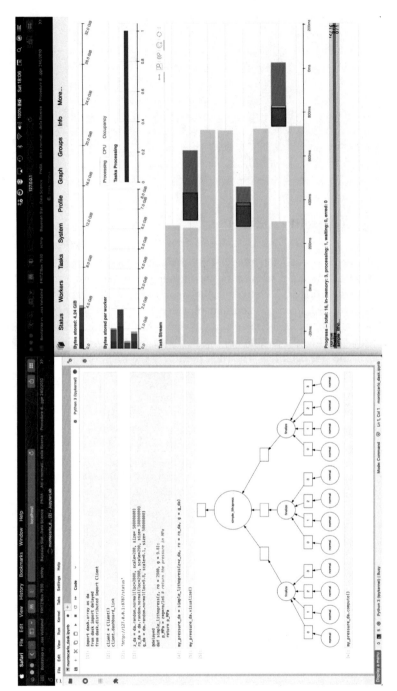

Fig. 10.18 Dask Interactive dashboard

References

Caesar Wu, R. B. (2015). *Cloud data centers and cost modeling: A complete guide to planning, designing and building a cloud data center*, (1st ed.). Burlington: Morgan Kaufmann.

Daniel, J. C. (2019). *Data science with python DASK*. New York: Manning Publications.

David, P. (2011). *Encyclopedia of parallel computing*. New York: Springer. https://doi.org/10.1007/978-0-387-09766-4

Fiore, M., & Devesas Campos, M. (2013). The algebra of directed acyclic graphs. In B. Coecke, L. Ong, & P. Panangaden (Eds.), *Computation, logic, games, and quantum foundations. The many facets of Samson Abramsky. Lecture notes in computer science* (Vol. 7860). Berlin, Heidelberg: Springer. https://doi.org/10.1007/978-3-642-38164-5_4

Maurer, S. B. (2013). *Directed acyclic graphs*. Routledge Handbooks Online. Milton Park: Routledge. https://doi.org/10.1201/B16132-10

Pedregosa, F., Varoquaux, G. G., Gramfort, A., Michel, V., Thirion, B., Grisel, O., Blondel, M., Prettenhofer, P., Weiss, R., Dubourg, V., Vanderplas, J., Passos, A., Cournapeau, D., Brucher, M., Perrot, M., & Duchesnay, É. (2011). Scikit-learn: Machine learning in python. *Journal of Machine Learning Research, 12*, 2825–2830.

Peter Pacheco, M. M. (2020). *An introduction to parallel programming* (2nd ed.). Burlington: Morgan Kaufmann.

Xu, J. (2003). *Theory and application of graphs* (vol. 10). New York: Springer. https://doi.org/10.1007/978-1-4419-8698-6

Chapter 11
Scale Your Models in the Cloud

11.1 Scaling Your Environment in the Cloud

The term "scalability" refers to the ability of a system to manage a growing amount of work. As stated in the previous chapter, compute or memory bounds must be scaled to handle ML models. In the context of a cloud computing facility, the term scaling refers to the ability to quickly and efficiently increase (or decrease) the capability of a computational resource to handle a model that no longer fits the current resources (i.e., RAM, CPUs, and storage capabilities). Two main strategies exist for scaling computational infrastructure: scale up or scale out (Bekkerman et al., 2012).

Scale Up
Scaling up, or vertical scaling, consists of replacing the current computational instance with something more powerful (Fig. 11.1). For example, we could increase the number of cores, the amount of memory, and/or the capability of the storage (Fig. 11.2).

Scale Out
Scaling out, or horizontal scaling, consists of increasing the computational capability by replicating the instances and running them in parallel (Fig. 11.3).

© The Author(s), under exclusive license to Springer Nature Switzerland AG 2023   177
M. Petrelli, *Machine Learning for Earth Sciences*, Springer Textbooks
in Earth Sciences, Geography and Environment,
https://doi.org/10.1007/978-3-031-35114-3_11

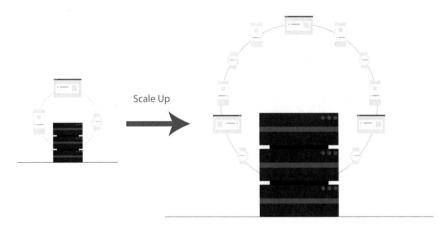

Fig. 11.1 Scaling up and scaling down

Compute Optimized Instances by Amazon Web Services

Instance Size	vCPU	Memory (GiB)	
r6a.large	2	16	Scale Down
r6a.xlarge	4	32	
r6a.2xlarge	8	64	
r6a.4xlarge	16	128	
r6a.8xlarge	32	256	
r6a.12xlarge	48	384	
r6a.16xlarge	64	512	
r6a.24xlarge	96	768	
r6a.32xlarge	128	1024	
r6a.48xlarge	192	1536	
r6a.metal	192	1536	

Scale Up

Fig. 11.2 Scaling up and down

11.2   Scaling in the Cloud: The Hard Way

The "hard way" of scaling consists of managing all configurations and taking all the technical steps in either Amazon Web Services (AWS), the Google Compute Engine, Microsoft Azure, or other providers.

Scaling up is quite easy with cloud providers. It consists simply of selecting larger or smaller instances to scale up and down, respectively (Fig. 11.2). Also, some providers offer specific services for auto-scaling; for example, Amazon claims that "AWS Auto Scaling monitors your applications and automatically adjusts capacity

Fig. 11.3 Scaling out

to maintain steady, predictable performance at the lowest possible cost. Using AWS Auto Scaling, it's easy to set up application scaling for multiple resources across multiple services in minutes. The service provides a simple, powerful user interface that lets you build scaling plans for resources including Amazon EC2 instances...."[1]

In contrast, scaling out is not as straightforward as scaling up. The Dask documentation suggests the use of Kubernetes and Helm solutions. Kubernetes is "a portable, extensible, open source platform for managing containerized workloads and services that facilitates both declarative configuration and automation."[2] Helm is "an open source package manager for Kubernetes. It provides the ability to provide, share, and use software built for Kubernetes."[3] The Dask documentation claims that "it is easy to launch a Dask cluster and a Jupyter notebook server on cloud resources using Kubernetes and Helm."[4] However, the instructions given in the Dask documentation assume that a Kubernetes cluster and Helm are already installed and ready for use. Unfortunately, setting up a Kubernetes cluster and Helm

---

[1] https://aws.amazon.com/autoscaling/.

[2] https://kubernetes.io/docs/concepts/overview/what-is-kubernetes/.

[3] https://helm.sh/docs/.

[4] https://docs.dask.org/en/stable/deploying-kubernetes-helm.html.

is not straightforward for a novice. Detailed instructions for many cloud providers are available in the guide "Zero to JupyterHub."[5]

11.3   Scaling in the Cloud: The Easy Way

Saturn Cloud

Saturn Cloud[6] is a cloud-based platform designed to support data scientists working with Python,[7] R,[8] Julia,[9] and other programming languages. Resources, such as those shown in Fig. 11.4, are the building blocks of the Saturn Cloud platform. The term "resource" refers to a complete computational and coding environment. Each resource is independent, so you can split out your different activities. Saturn Cloud-hosted solutions[10] are a "pay as you go" service, which means that you pay per hour for computational resources. For example, during the writing of the present book, the Medium (2 vCPU and 4 GB of RAM) and V100-16×Large (64 vCPU, 8 vGPU, and 488 GB of RAM) resources cost \$0.06 and \$34.24 per hour, respectively. A free hosted plan also exists with limited resources. The next sections exploit the free hosted plan for the first step of scaling up, following which results obtained on a Hosted Pro Plan are shown. Details about the costs are also provided, in case you intend to reproduce these results.

Speed Up GridSearchCV on Saturn Cloud

In Sect. 8.3, we performed a GridSearchCV, which is an extensive search within the hyper-parameters governing the extremely randomized trees algorithm (see code listing 8.9 and Table 8.1). The aim was to find the combination of hyper-parameters that provide the highest degree of accuracy. This combination of hyper-parameters resulted in a grid of 48 models, each repeated three times through cross validation, for a total of 144 attempts. As reported in Chap. 8, running the code listing 8.9 required about 8 hours on my MacBook pro equipped with a 2.3 GHz Quad-Core Intel™ i7 CPU and 32 GB of RAM.

---

[5] https://zero-to-jupyterhub.readthedocs.io/en/latest/kubernetes/.

[6] https://saturncloud.io.

[7] https://www.python.org.

[8] https://www.r-project.org.

[9] https://julialang.org.

[10] https://saturncloud.io/plans/hosted/.

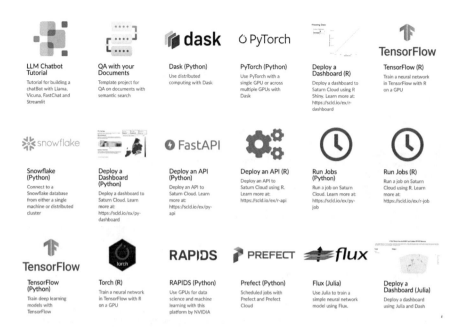

Fig. 11.4   Saturn cloud computing templates

In Saturn Cloud, the free hosted plan allows a slight scaling up of the hardware that supports my MacBook Pro using the $2\times$ Large instance (i.e., eight cores and 64 GB of RAM), so we scale up to a $2\times$ Large instance and run the code listing 8.9. To start, we register with Saturn Cloud and click on the "New Python Server" button (Fig. 11.5), which starts a guided procedure that allows the configuration of a new instance, ready for basic Python data analysis, machine learning, and, possibly, parallel processing with Dask.

Figures 11.6 and 11.7 show all the steps to configure the new instance. It is recommended to use a self-explanatory name, such as *scale_GridSearchCV_Joblib*, 100 Gi of disk space, and the $2\times$large instance. Also, remember to add *ytables* as an extra package; this is installed using *Conda Install*. The PyTables library allows HDF5 files to be read and saved. Leave all the other options untouched, and click *Create*.

The instance is now ready (Fig. 11.8). The next steps consist of starting the instance, creating a new Jupyter Notebook, and uploading the HDF5 file *ml_data.h5* (Fig. 11.9). Finally, we are ready to replicate code listing 8.9 in a $2\times$large instance (Fig. 11.10). Note that the second block of code in (Fig. 11.10) simply reports the outputs in a log file named *data.log*. The fitting (i.e., block number five) lasted 5 hours and 15 minutes, which is significantly faster than the 8 hours of my MacBook Pro.

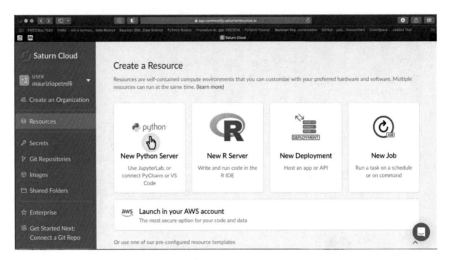

Fig. 11.5 Starting a new python server

Next, we activate a Hosted Pro Plan[11] and progressively scale up the code in
Fig. 11.10 to 8×Large instances (i.e., 32 Cores and 256 GB of RAM at the cost
of $3.30/hour) and 16×Large instances (i.e., 64 Cores and 512 GB of RAM at the
cost of $6.59/hour), improving the computation time to about 2 hours and 1 hour,
respectively.

As a final step, I scaled out the code reported in Fig. 11.10. To do this, I created
a Dask cluster by clicking *New Dask Cluster* (Fig. 11.8), which opens the Cluster
configuration window (Fig. 11.11). Also, I opted for a 16×Large scheduler (i.e.,
64 Cores and 512 GB of RAM) and four 8×Large workers (i.e., 32 Cores and
256 GB of RAM). To run *GridSearchCV* in the newly created Dask Cluster, the
code reported in Fig. 11.10 requires only minimal changes, which are all reported in
Fig. 11.12. I imported *SaturnCluster* from *dask_saturn* (block 1), used $n\_jobs = -1$
(i.e., nested parallelism) for both *ExtraTreesClassifier* and *GridSearchCV* (Block 4),
defined the *SaturnCluster* client (Block 5), and ran *Joblib* with *dask* as the fitting
engine (Block 6). In this final case, fitting *GridSearchCV* required less than 25
minutes!

---

[11] https://saturncloud.io/plans/hosted/.

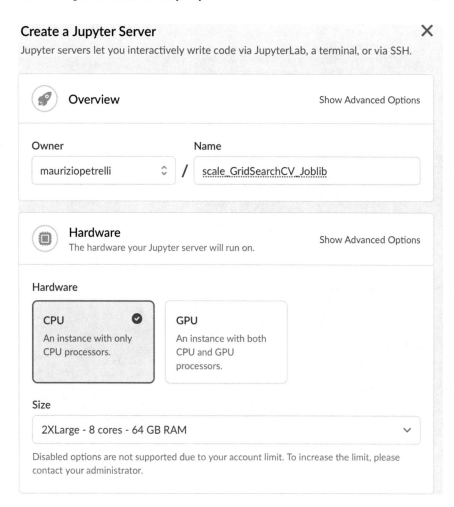

Fig. 11.6 Setting up the python server parameters

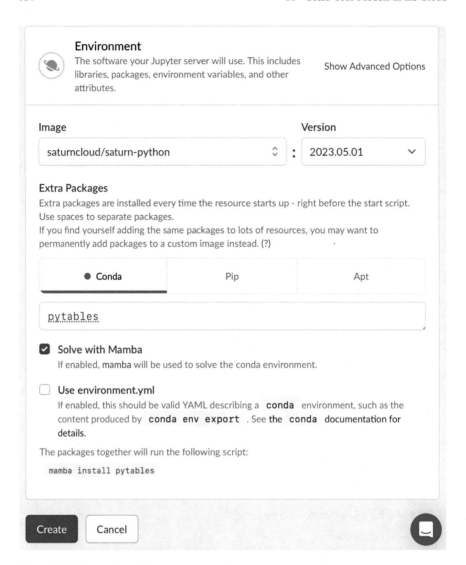

Fig. 11.7 Setting up the python server parameters

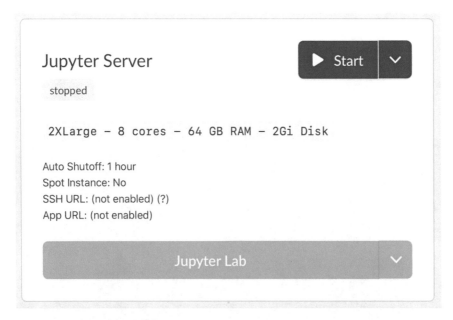

Fig. 11.8 Starting the python server

Fig. 11.9 Uploading a hdf5 file

```
[1]: import joblib as jb
     import pandas as pd
     from sklearn.ensemble import ExtraTreesClassifier
     from sklearn.model_selection import GridSearchCV, train_test_split
     from sklearn.pipeline import Pipeline
     from sklearn.preprocessing import StandardScaler
     Last executed at 2023-08-03 17:36:30 in 660ms
```

```
[2]: import logging
     import sys

     so = open("data.log", 'w', 10)
     sys.stdout.echo = so
     sys.stderr.echo = so

     get_ipython().log.handlers[0].stream = so
     get_ipython().log.setLevel(logging.INFO)
     Last executed at 2023-08-03 17:36:31 in 4ms
```

```
[3]: X = pd.read_hdf('~/workspace/ml_data.h5', 'train').values
     y = pd.read_hdf('~/workspace/ml_data.h5', 'train_target').values

     X_train, X_test, y_train, y_test = train_test_split(
         X, y, test_size=0.2, random_state=10, stratify=y)
     Last executed at 2023-08-03 17:36:36 in 682ms
```

```
[4]: param_grid = {
         'classifier__criterion': ['entropy', 'gini'],
         'classifier__min_samples_split': [2, 5, 8, 10],
         'classifier__max_features': ['sqrt', 'log2', None],
         'classifier__class_weight': ['balanced', None]
         }

     clf = Pipeline([('scaler',  StandardScaler()),
                 ('classifier',ExtraTreesClassifier(n_estimators=250, n_jobs=-1))])

     CV_rfc = GridSearchCV(estimator=clf, refit=True, param_grid=param_grid,
                           cv= 3, verbose=10)
     Last executed at 2023-08-03 17:36:36 in 4ms
```

```
[5]: CV_rfc.fit(X_train, y_train)
```

```
[6]: jb.dump(CV_rfc, 'ETC_grid_search_results_rev_3_baseline.pkl')
     Last executed at 2023-08-03 18:24:12 in 4.99s
```

```
[6]: ['ETC_grid_search_results_rev_3_baseline.pkl']
```

```
[7]: CV_rfc.best_params_
     Last executed at 2023-08-03 18:24:13 in 4ms
```

```
[7]: {'classifier__class_weight': 'balanced',
      'classifier__criterion': 'entropy',
      'classifier__max_features': None,
      'classifier__min_samples_split': 2}
```

Fig. 11.10 Scaling Up the *GridSearchCV*

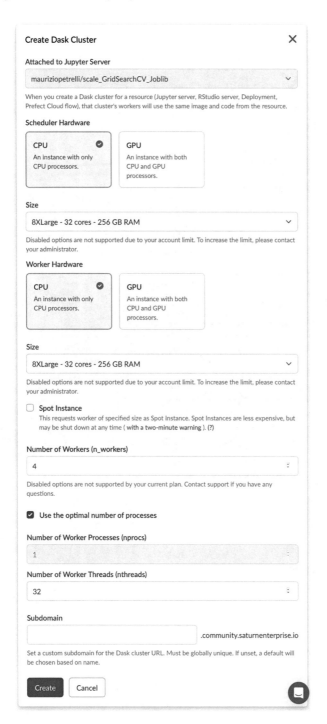

Fig. 11.11 Setting up a new dask cluster

```
[1]:  import joblib as jb
      import pandas as pd
      from sklearn.ensemble import ExtraTreesClassifier
      from sklearn.model_selection import GridSearchCV, train_test_split
      from sklearn.pipeline import Pipeline
      from sklearn.preprocessing import StandardScaler
      from dask_saturn import SaturnCluster
      from dask.distributed import Client
      Last executed at 2023-08-03 17:03:38 in 880ms
```

```
[2]:  import logging
      import sys

      so = open("data.log", 'w', 10)
      sys.stdout.echo = so
      sys.stderr.echo = so

      get_ipython().log.handlers[0].stream = so
      get_ipython().log.setLevel(logging.INFO)
      Last executed at 2023-08-03 17:03:39 in 4ms
```

```
[3]:  X = pd.read_hdf('~/workspace/ml_data.h5', 'train').values
      y = pd.read_hdf('~/workspace/ml_data.h5', 'train_target').values

      X_train, X_test, y_train, y_test = train_test_split(
          X, y, test_size=0.2, random_state=10, stratify=y)
      Last executed at 2023-08-03 17:03:41 in 791ms
```

```
[4]:  param_grid = {
          'classifier__criterion': ['entropy', 'gini'],
          'classifier__min_samples_split': [2, 5, 8, 10],
          'classifier__max_features': ['sqrt', 'log2', None],
          'classifier__class_weight': ['balanced', None]
          }

      clf = Pipeline([('scaler',  StandardScaler()),
                  ('classifier',ExtraTreesClassifier(n_estimators=250, n_jobs=-1))])

      CV_rfc = GridSearchCV(estimator=clf, refit=True, param_grid=param_grid,
                              cv= 3, verbose=1, n_jobs=-1)
      Last executed at 2023-08-03 17:03:42 in 5ms
```

```
[5]:  client = Client(SaturnCluster())
      Last executed at 2023-08-03 17:03:45 in 660ms

      INFO:dask-saturn:Cluster is ready
      INFO:dask-saturn:Registering default plugins
      INFO:dask-saturn:Success!
```

```
[6]:  with jb.parallel_backend("dask"):
          _ = CV_rfc.fit(X_train, y_train)
      Last executed at 2023-08-03 17:27:58 in 24m 6.78s
      Fitting 3 folds for each of 48 candidates, totalling 144 fits
```

```
[8]:  jb.dump(_, 'ETC_grid_search_results_rev_3_dask.pkl')
      Last executed at 2023-08-03 17:28:33 in 4.95s
```

```
[8]:  ['ETC_grid_search_results_rev_3_dask.pkl']
```

```
[9]:  _.best_params_
      Last executed at 2023-08-03 17:28:38 in 5ms
```

```
[9]:  {'classifier__class_weight': 'balanced',
       'classifier__criterion': 'entropy',
       'classifier__max_features': None,
       'classifier__min_samples_split': 2}
```

Fig. 11.12 Scaling out *GridSearchCV*

Reference

Bekkerman, R., Bilenko, M., & Langford, J. (2012). *Scaling up machine learning: Parallel and distributed approaches*. Cambridge: Cambridge University Press.

Part V
Next Step: Deep Learning

Chapter 12
Introduction to Deep Learning

12.1 What Does Deep Learning Mean?

As introduced in Chap. 1, ML algorithms gather knowledge by extracting patterns from data.

In other words, they try to map the representation provided by the investigated features to produce an output (Goodfellow et al., 2016). Therefore, features are central in ML because they provide the information to build a representation. However, simply mapping a representation to deliver an output is often insufficient. Therefore, we must train ML systems to discover not only the mapping from representation to output but also the representation itself (Goodfellow et al., 2016). This approach is known as representation learning. In complex problems (e.g., problems characterized by many features or extremely large data sets), learning the representation is not straightforward.

"Deep learning solves this central problem in representation learning by introducing representations that are expressed in terms of other, simpler representations. Deep learning enables the computer to build complex concepts out of simpler concepts" (Goodfellow et al., 2016).

A typical example of deep learning is the multilayer perceptron, which is a mathematical function that maps a set of inputs to output values (Goodfellow et al., 2016). The function is formed by combining many simpler functions (Fig. 12.1). To better understand, Fig. 12.1 shows how a deep learning method can represent the

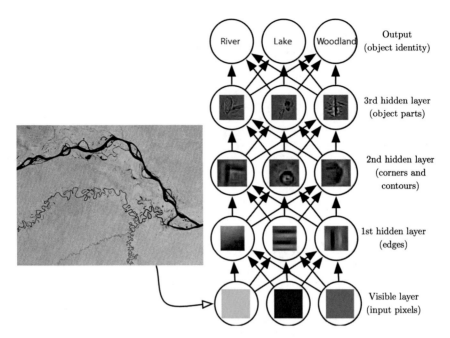

Fig. 12.1 Illustration of a deep learning, multilayer perceptron model. Modified from Goodfellow et al. (2016). The image comes from Copernicus Sentinel-1 mission and shows the meandering Amazon River (https://www.esa.int/ESA_Multimedia/Images/2020/09/Amazon_River)

concept of an image by combining simpler notions, such as corners and contours, which are in turn defined in terms of edges (Goodfellow et al., 2016). In Fig. 12.1, the input feeds the visible layer and then a series of hidden layers progressively extracts and elaborates abstract features from the initial inputs. The final layer provides the output (e.g., the result of mapping the representation developed during the learning process) (Goodfellow et al., 2016).

From the mathematical point of view, a deep feedforward network (or multilayer perceptron) aims to approximate some function f^* (Goodfellow et al., 2016). In detail, it defines a mapping $\mathbf{y} = f(\mathbf{x}; \boldsymbol{\theta})$ and learns the value of the parameters $\boldsymbol{\theta}$ that result in the most accurate approximation of the function (Goodfellow et al., 2016) (Fig. 12.2). Why feedforward? Because data flow through the function from the input $\mathbf{x}$, through the intermediate computations used to define f, and finally to the output $\mathbf{y}$. Why networks? Because networks are typically expressed by combining many different functions. For example, we might combine three functions $f^{(1)}$, $f^{(2)}$, and $f^{(3)}$ in a chain to define $f(x) = f^{(3)}(f^{(2)}(f^{(1)}(x)))$ (Goodfellow et al., 2016). In detail, $f^{(1)}$ is the first layer of the network, $f^{(2)}$ is the second layer, and so on (Goodfellow et al., 2016). The overall length of the chain defines the depth of the model. That's why they are deep. The final layer of a feedforward network provides the output. During the training process, we adjust $\boldsymbol{\theta}$ parameters in $f(\mathbf{x}; \boldsymbol{\theta})$ to match $f^*(\mathbf{x})$ (Goodfellow et al., 2016).

Fig. 12.2 Example of three-layer feedforward network or multilayer perceptron

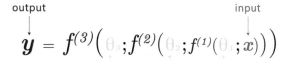

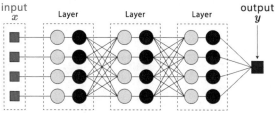

Fig. 12.3 Vectors, matrices, tensors

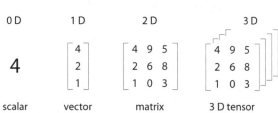

12.2 PyTorch

"PyTorch is an optimized tensor library for deep learning using GPUs and CPUs."[1] Tensors (i.e., multidimensional arrays) are at the base of PyTorch. Also, PyTorch hosts the *autograd* engine (see *torch.autograd*), which can compute derivatives, even providing complex data structures. The other PyTorch modules are mainly based on tensors and on the *autograd* engine. For example, the *torch.nn* module provides common neural network layers and other architectural components. The *torch.optim* implements *state-of-the-art* optimization strategies for the learning process (Imambi et al., 2021).

12.3 PyTorch Tensors

PyTorch tensors are multidimensional arrays (Fig. 12.3), similar to those in NumPy. However, in contrast with NumPy arrays, PyTorch tensors can (1) perform accelerated operations on graphical processing units (GPUs), (2) natively work on distributed environments, and (3) keep track of a graph of operations when necessary (Imambi et al., 2021). The initialization of PyTorch tensors mimics what is done with NumPy arrays. Finally, Numpy arrays can be easily imported as PyTorch tensors (Fig. 12.4).

---

[1] https://pytorch.org/docs/stable/index.html.

```
[1]: import torch
```
Last executed at 2022-08-15 09:51:44 in 7.11s

1D tensor → vector

```
[2]: zeros = torch.zeros(6)
     print(zeros)
```
Last executed at 2022-08-15 09:40:55 in 84ms

```
tensor([0., 0., 0., 0., 0., 0.])
```

2D tensor → matrix

```
[3]: ones = torch.ones(2, 3)
     print(ones)
```
Last executed at 2022-08-15 09:40:56 in 27ms

```
tensor([[1., 1., 1.],
        [1., 1., 1.]])
```

3D tensor

```
[4]: random1 = torch.randn(2, 3, 3)
     print(random1)
```
Last executed at 2022-08-15 09:40:57 in 8ms

```
tensor([[[ 0.2335, -1.5447,  1.4459],
         [ 0.9062,  0.9170,  2.1128],
         [-0.8420,  0.0852,  1.4125]],

        [[ 2.5360,  0.0471, -0.7606],
         [ 0.3327,  0.4224,  0.4928],
         [-0.3284, -0.5549,  0.2324]]])
```

Tensor from a Python list

```
[5]: my_list = [3, 6, 8, 6, 8, 9]
     tensor1 = torch.tensor(my_list, dtype = torch.float)
     print(tensor1)
```
Last executed at 2022-08-15 09:40:58 in 5ms

```
tensor([3., 6., 8., 6., 8., 9.])
```

Tensor from NumPy array

```
[6]: import numpy as np

     np_array = np.array([3, 4, 5, 7, 9])
     tensor2 = torch.from_numpy(np_array)
     tensor3 = torch.tensor(np_array)
     print(tensor2)
     print(tensor3)
```
Last executed at 2022-08-15 09:41:02 in 4ms

```
tensor([3, 4, 5, 7, 9])
tensor([3, 4, 5, 7, 9])
```

Fig. 12.4 Vectors, matrices, tensors

```
[1]: import torch
     Last executed at 2022-08-17 16:45:10 in 7.27s
```

Working on the GPU

```
[2]: torch.cuda.is_available()
     Last executed at 2022-08-17 16:45:10 in 34ms
```

```
[2]: True
```

```
[3]: random_cuda1 = torch.randn(500000000, device='cuda')
     random_cuda2 = torch.randn(500000000, device='cuda')
     Last executed at 2022-08-17 16:45:17 in 7.31s
```

```
[4]: power_cuda = random_cuda1 ** random_cuda2
     Last executed at 2022-08-17 16:45:17 in 13ms
```

```
[5]: random_cpu1 = torch.randn(500000000, device='cpu')
     random_cpu2 = torch.randn(500000000, device='cpu')
     Last executed at 2022-08-17 16:45:26 in 8.55s
```

```
[6]: power_cpu = random_cpu1 ** random_cpu2
     Last executed at 2022-08-17 16:45:29 in 3.10s
```

Fig. 12.5 Vectors, matrices, tensors

By default, PyTorch tensors live on the CPU. However, they can be easily defined on the GPU, if available (see block 2 of Fig. 12.5), by using the *device* parameter (i.e., *device='cuda'*, block 3 of Fig. 12.5). Blocks 3–6 in Fig. 12.5 simply highlight that the power operation performed on the *'cuda'* device (i.e., the GPU) lasts only 7 ms, which is much faster than the ≈ 3 s required to execute the same operation on the CPU.

12.4 Structuring a Feedforward Network in PyTorch

Figure 12.6 shows how to develop in PyTorch the feedforward neural network (i.e., a multilayer perceptron) shown in Fig. 12.2.

The feedforward neural network consists of an input layer (layer 1) that accepts input vectors with four features. ReLu functions process the input features and forward the results to a hidden layer (layer 2), which is characterized by four neurons and a ReLu activation function (i.e., the ReLu function). Finally, the output layer returns a scalar as output.

In PyTorch, a neural network is a module with a nested structure. In other words, a neural network consists of a module that contains other modules (i.e., layers). The

```
[1]:  import torch
      from torch import nn
      Last executed at 2022-08-17 16:34:00 in 571ms
```

```
[2]:  class MultilayerPerceptron(nn.Module):
          '''
          Example of Multilayer Perceptron
          '''
          def __init__(self):
              super().__init__()
              self.layers = nn.Sequential(
                nn.Linear(4, 4),
                nn.ReLU(),
                nn.Linear(4, 4),
                nn.ReLU(),
                nn.Linear(4, 1)
                )

          def forward(self, x):
              return self.layers(x)
      Last executed at 2022-08-17 16:34:00 in 4ms
```

```
[3]:  device = "cuda" if torch.cuda.is_available() else "cpu"

      print(f"Using {device} device")
      Last executed at 2022-08-17 16:34:00 in 4ms
      Using cpu device
```

```
[4]:  model = MultilayerPerceptron().to(device)

      print(model)
      Last executed at 2022-08-17 16:34:00 in 32ms
      MultilayerPerceptron(
        (layers): Sequential(
          (0): Linear(in_features=4, out_features=4, bias=True)
          (1): ReLU()
          (2): Linear(in_features=4, out_features=4, bias=True)
          (3): ReLU()
          (4): Linear(in_features=4, out_features=1, bias=True)
        )
      )
```

Fig. 12.6 Developing a multilayer perceptron in PyTorch

model can live either in the CPU or in the GPU (Blocks 3 and 4 in Fig. 12.6), if available.

12.5   How to Train a Feedforward Network

12.5.1   The Universal Approximation Theorem

The universal approximation theorem (Hornik et al., 1989; Cybenko, 1989) states that feedforward networks with a linear output layer and at least one hidden layer can approximate any continuous function on a closed and bounded subset of $\mathbb{R}^n$ (Goodfellow et al., 2016), which means that feedforward networks with hidden layers are universal approximators (Goodfellow et al., 2016). In other words, "the universal approximation theorem means that regardless of what function we are trying to learn, we know that a large [multilayer perceptron] will be able to represent this function" (Goodfellow et al., 2016). However, despite what is affirmed by the universal approximation theorem, there is no guarantee that the training process will correctly learn the target function (Goodfellow et al., 2016). For example, the optimization algorithm used for training may not be able to find the correct values for the theta parameters that describe the desired function. Also, the training process might choose the wrong function because of overfitting (Goodfellow et al., 2016). To avoid these issues, we want to find (1) a robust loss function $L(\theta)$, (2) a strategy to compute the gradient with respect to model parameters [i.e., $\Delta_\theta L(\theta)$ of $L(\theta)$], and (3) an efficient optimization algorithm to descend $\Delta_\theta L(\theta)$ and find the minimum of $L(\theta)$.

12.5.2 Loss Functions in PyTorch

A loss function (or cost function) computes a numerical value that the learning process will attempt to minimize (cf. Sect. 7.5). Typically, a loss function compares (e.g., by subtraction) the desired outputs (i.e., the labels) and the current outputs of our model (Stevens et al., 2020). Table 12.1 reports the loss functions available in PyTorch.

12.5.3 The Back-Propagation and its Implementation in PyTorch

In feedforward neural networks, the information starts from the input $\mathbf{x}$, flows through the hidden layers, and finally produces an output $\mathbf{y}$ (Goodfellow et al., 2016). The name of this process is forward propagation. At the beginning of training, forward propagation produces an output $\mathbf{y}$ and an associated cost function $J(\theta)$ that relies on the non-optimized θ parameters (Goodfellow et al., 2016).

The back-propagation algorithm computes the gradient of $L(\theta)$ by propagating the information from the output (i.e., the cost function), backward through the network (Goodfellow et al., 2016). Note that back-propagation only allows us to define the gradient of $L(\theta)$. We then need an optimization algorithm such as the stochastic gradient descent algorithm (Sect. 7.5) to learn along this gradient (Goodfellow et al., 2016). Describing in detail the back-propagation algorithm is beyond the scope of the present book, so please refer to Goodfellow et al. (2016) or other specialized books for further details.

Table 12.1 Loss functions in PyTorch: https://bit.ly/pyt-loss-functions

Loss function	Description
nn.L1Loss	Loss function based on mean absolute error (MAE)
nn.MSELoss	Loss function based on mean squared error (squared L2 norm)
nn.CrossEntropyLoss	Computes cross entropy loss between input and target
nn.CTCLoss	Connectionist temporal classification loss
nn.NLLLoss	Negative log likelihood loss
nn.PoissonNLLLoss	Negative log likelihood loss with Poisson distribution of target
nn.GaussianNLLLoss	Gaussian negative log likelihood loss
nn.KLDivLoss	Kullback–Leibler divergence loss
nn.BCELoss	Binary cross entropy between target and input probabilities
nn.BCEWithLogitsLoss	Combines Sigmoid layer and BCELoss in one single class
nn.MarginRankingLoss	Measures the loss given inputs x_1, x_2, two one-dimensional mini-batch or zero-dimensional tensors, and a label one-dimensional mini-batch or zero-dimensional tensor y (containing 1 or -1)
nn.HingeEmbeddingLoss	Masures loss given an input tensor x and a labels tensor y (containing 1 or -1)
nn.MultiLabelMarginLoss	Optimizes a multi-class multi-classification hinge loss (margin-based loss)
nn.HuberLoss	Creates a criterion that uses a squared term if the absolute element-wise error falls below delta and a delta-scaled L1 term otherwise (Huber loss).
nn.SmoothL1Loss	Creates a criterion that uses a squared term if the absolute element-wise error falls below beta and an L1 term otherwise
nn.SoftMarginLoss	Creates a criterion that optimizes a two-class classification logistic loss between input tensor x and target tensor y (containing 1 or -1)
nn.MultiLabelSoftMarginLoss	Optimizes a multi-label one-versus-all loss based on max-entropy, between input x and target y of size (N, C).
nn.CosineEmbeddingLoss	Measures the loss given input tensors x_1, x_2 and a tensor label y with values 1 or -1.
nn.MultiMarginLoss	Creates and optimizes a multi-class classification hinge loss (margin-based loss)
nn.TripletMarginLoss	Measures the triplet loss given input tensors x_1, x_2, x_3 and a margin with a value greater than zero
nn.TripletMarginWithDistanceLoss	Measures triplet loss given input tensors a, p, and n (representing anchor, positive, and negative examples, respectively), and a nonnegative, real-valued function ("distance function") used to compute the relationship between the anchor and a positive example ("positive distance") and between the anchor and a negative example ("negative distance")

Table 12.2 Optimization algorithms in PyTorch: https://bit.ly/pytorch-optim

Optimization algorithm	Description
Adadelta	Implements Adadelta algorithm
Adagrad	Implements Adagrad algorithm
Adam	Implements Adam algorithm
AdamW	Implements AdamW algorithm
SparseAdam	Implements lazy version of Adam algorithm suitable for sparse tensors
Adamax	Implements Adamax algorithm (a variant of Adam based on infinity norm)
ASGD	Implements averaged stochastic gradient descent
LBFGS	Implements L-BFGS algorithm, heavily inspired by minFunc
NAdam	Implements NAdam algorithm
RAdam	Implements RAdam algorithm
RMSprop	Implements RMSprop algorithm
Rprop	Implements the resilient backpropagation algorithm
SGD	Implements stochastic gradient descent (optionally with momentum)

The engine *torch.autograd* is PyTorch's automatic differentiation engine. It defines a directed acyclic graph whose leaves are the input tensors and whose roots are the output tensors. In this way, it computes gradients via the chain rule.

12.5.4   Optimization

Once defined, the optim submodule of torch (i.e. *torch.optim*) stores the optimization algorithms (Table 12.2).

12.5.5   Network Architectures

This section provides a quick overview of some popular neural network architectures.

Multilayer Perceptron

A multilayer perceptron is the neural network structure depicted in Fig. 12.2. It consists of fully connected layers of perceptrons (i.e., artificial neurons). Selecting the optimal number of hidden layers is not always straightforward and is commonly driven by background knowledge and experimentation (Hastie et al., 2017). With too few hidden units, the model might not have enough flexibility to capture the nonlinearities in the data; with too many hidden units, the extra weights can be shrunk toward zero if appropriate regularization is used." Common applications typically use 5–100 hidden layers (Hastie et al., 2017). Most ML models described in Chap. 7 (e.g., support vector machines or logistic regression) can be simulated by multilayer perceptrons containing only one or two layers (Aggarwal, 2018).

Radial Basis Function Networks

Radial basis function networks consist of shallow (i.e., only two layers) neural networks where the first and the second layers are unsupervised and supervised, respectively (Aggarwal, 2018). Radial basis function networks are based on Cover's theorem on the separability of patterns (Cover, 1965), stating that pattern classification problems are more likely to be linearly separable when cast into a high-dimensional space with a nonlinear transformation. The idea behind radial basis function networks is close to that of nearest-neighbor classifiers with the addition of a supervised step in the second layer (Aggarwal, 2018). Also, they are similar to support vector machines trained with radial basis functions as the kernel. However, radial basis function networks are more general than kernel support vector machines (Aggarwal, 2018).

Restricted Boltzmann Machines

Restricted Boltzmann machines (RBMs) are unsupervised neural network architectures that rely on energy minimization (Fischer & Igel, 2012). Although RBMs were introduced in the 1980s (Aggarwal, 2018), the increase in computational power and the development of new learning strategies has made RBMs significantly more appealing in recent years (Fischer & Igel, 2012). RBMs are useful for creating generative models (Fischer & Igel, 2012) and are closely related to probabilistic graphical models (Koller & Friedman, 2009). Also, RBMs have been proposed as building blocks for so-called "deep belief networks" ((Hinton et al., 2006). Training a RBM is rather different from training a feedforward network because it cannot use backpropagation (Fischer & Igel, 2012). On the contrary, RBMs rely on Monte Carlo sampling for the training (Fischer & Igel, 2012).

Recurrent Neural Networks

Recurrent neural networks (RNNs) are designed to investigate sequential data such as text sentences, time series, and other discrete sequences (Abraham and Tyagi, 2022). An important point about RNNs is that they account for the potential dependence of subsequent inputs on previous inputs, making them well suited, for example, for time series forecasting or speech recognition (Kumar & Abraham, 2022; Aggarwal, 2018). RNNs use a specific backpropagation algorithm called "backpropagation through time" (Aggarwal, 2018), which accounts for the sequential nature of the inputs during the learning process. A drawback of RNNs is their complex optimization and training processes, making them difficult to access, especially for novices (Kumar & Abraham, 2022; Aggarwal, 2018). Specialized variants of the recurrent neural network architecture have also been proposed to solve specific problems, such as handling long-term dependencies using long short-term memory networks (Hochreiter & Schmidhuber, 1997)

Convolutional Neural Networks

Convolutional neural networks (CNNs) are biologically inspired networks that find applications in video and speech recognition, recommendation systems, image classification and segmentation, natural language processing, and time series forecasting (see, e.g., Yamashita et al., 2018). CNNs mimic the visual cortex functionalities of animals (Fukushima, 1980) and aim to "automatically and adaptively learn spatial hierarchies of features through backpropagation by using multiple building blocks, such as convolution layers, pooling layers, and fully connected layers" (Fukushima, 1980).

CNNs are well suited to process grid-shaped data such as RGB images or spectral maps by using three main types of layers: convolution, pooling, and fully connected (Fukushima, 1980). The first two layer types extract features and the third layer maps the extracted features to the final output.

Convolution layers play a fundamental role in CNNs (Yamashita et al., 2018). They typically consist of three components: input data, a filter (or kernel), and a feature map (Yamashita et al., 2018). To better understand, consider the example shown in Fig. 12.7, where the input and the kernel are 6×6 and 3×3 arrays, respectively. The output is a 4×4 array named "feature map," "activation map," or "convolved feature" and derives from the systematic application of the filter (i.e., a dot product) to different portions of the input. After each convolution, the CNN applies an activation function such as a rectified linear unit (ReLU) to the output and then moves to the next layer (Yamashita et al., 2018).

Pooling layers reduce the dimensionality (or downsample), which reduces the number of parameters in the input. They typically consist of a filter that applies an aggregation function such as the max or average pooling (Fukushima, 1980). Max pooling selects the pixel with the maximum output of the filter and sends it to the output array. Similarly, average pooling calculates the average value within the filter and sends it to the output array. If you complain that a huge amount of information is lost in the pooling layers, you would be right. However, pooling layers reduce the complexity of the model, improve its efficiency, and limit the risk of overfitting (Fukushima, 1980). Finally, fully connected layers mimic a multilayer perceptron. For example, CNNs are widely used in semantic image segmentation (see, e.g., Badrinarayanan et al., 2017; Long et al., 2015; Milletari et al., 2016). Semantic image segmentation consists of identifying the areas (i.e., the pixels) of the image occupied by a specific subject, such as a person, as in the case of Fig. 12.8.

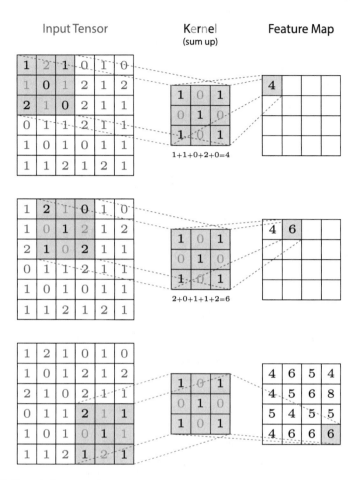

Fig. 12.7 Convolution example

12.6   Example Application

The Problem

As an example application of deep learning potentials in the Earth Sciences, we now discuss the training and validation of a CNN to identify building footprints from satellite records.

The problem falls in the ML classification sub-field called "semantic image segmentation" (see Fig. 12.8). In this specific case, we want to identify the areas or the pixels of an image occupied by buildings in the aerial image labeling data set (Maggiori et al., 2017) (see, e.g., Fig. 12.9). The right panel of Fig. 12.9 shows the solution to the problem in the form of a mask where white and black define

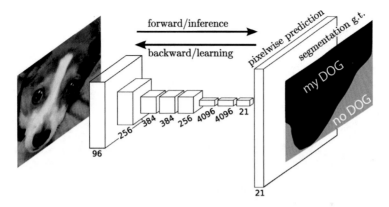

Fig. 12.8 Convolutional neural networks for image segmentation. Modified from Long et al., 2015

Fig. 12.9 The aerial image labeling data set (Maggiori et al., 2017)

building and non-building areas, respectively. We want to know whether we can train a CNN to produce the solution reported in Fig. 12.9. To attempt a simplified solution, I trained the U-Net CNN (Ronneberger et al., 2015) using PyTorch.

Data Set and Pre-processing
As a starting point, I downloaded the aerial image labeling data set (Maggiori et al., 2017), which consists of 360 orthorectified RGB (Red, Green, Blue) images linked to official cadastral records (Maggiori et al., 2017). The entire data set covers several areas, such as Austin (USA), Chicago (USA), Vienna (Austria), East and West Tyrol (Austria), San Francisco (USA), and Innsbruck (Austria). The lateral resolution is 0.3 m, and each tile is 5000 × 5000 pixels (Maggiori et al., 2017). For 180 tiles, a mask containing two semantic classes, building and non-building, is also provided (Maggiori et al., 2017). For the case study provided herein, I selected 10 tiles from

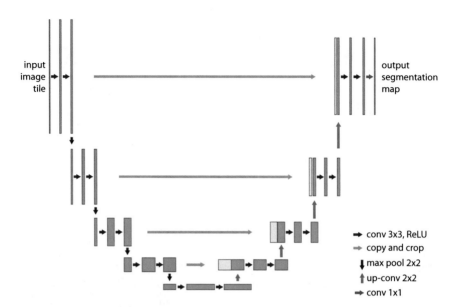

Fig. 12.10 Architecture of a U-net convolutional neural network (modified from Ronneberger et al., 2015)

Austin. For each tile, I also collected the associated masks to train and validate the model. From each tile, I extracted 25 images of 1000 × 1000 pixels each by using a 5 × 5 grid (the same operation was done for each mask). The resulting data set consisted of 245 images and 245 masks. I then split the data set into two parts for use in training (220) and validation (25).

The U-Net Architecture
The U-Net is a "fully convolutional network" (Long et al., 2015). The main concept behind fully convolutional networks is to take an input of arbitrary size and produce a correspondingly sized output with efficient inference and learning (Long et al., 2015).

Figure 12.10 shows the U-Net architecture. It consists of a contracting network (left side) followed by an expansive path (right side; Ronneberger et al., 2015). The contracting path applies a sequence of two 3 × 3 convolutions, each followed by a ReLU and 2 × 2 max pooling (Ronneberger et al., 2015). Next, in the expansive path, the U-net architecture upsamples the feature map, followed by a 2 × 2 convolution ("up-convolution"), and two 3 × 3 convolutions, each followed by a ReLU (Ronneberger et al., 2015). The final layer applies a 1 × 1 convolution to map each 64-component feature vector to the desired number of classes (Ronneberger et al., 2015). The code listing 12.1 shows a PyTorch implementation of the U-net.[2]

```
1  """ Full assembly of the parts to form the complete network """
2
3  from .unet_parts import *
4
5
6  class UNet(nn.Module):
7      def __init__(self, n_channels, n_classes, bilinear=False):
8          super(UNet, self).__init__()
9          self.n_channels = n_channels
10         self.n_classes = n_classes
11         self.bilinear = bilinear
12
13         self.inc = DoubleConv(n_channels, 64)
14         self.down1 = Down(64, 128)
15         self.down2 = Down(128, 256)
16         self.down3 = Down(256, 512)
17         factor = 2 if bilinear else 1
18         self.down4 = Down(512, 1024 // factor)
19         self.up1 = Up(1024, 512 // factor, bilinear)
20         self.up2 = Up(512, 256 // factor, bilinear)
21         self.up3 = Up(256, 128 // factor, bilinear)
22         self.up4 = Up(128, 64, bilinear)
23         self.outc = OutConv(64, n_classes)
24
25     def forward(self, x):
26         x1 = self.inc(x)
27         x2 = self.down1(x1)
28         x3 = self.down2(x2)
29         x4 = self.down3(x3)
30         x5 = self.down4(x4)
31         x = self.up1(x5, x4)
32         x = self.up2(x, x3)
33         x = self.up3(x, x2)
34         x = self.up4(x, x1)
35         logits = self.outc(x)
36         return logits
```

Listing 12.1 U-Net implementation in PyTorch

Results

Figure 12.11 shows the result of applying the trained model (1260 epochs) to one of the 25 validation images extracted from the original data set. The top-right panel shows the original image (i.e., the input RGB matrix), and the top-left panel shows the building–non-building mask. Keep in mind that we used building–non-building

---

[2] https://github.com/milesial/Pytorch-UNet.

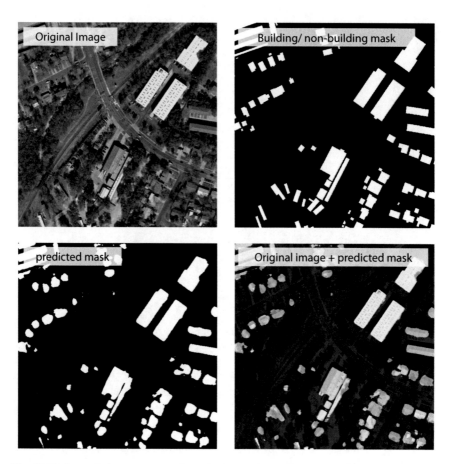

Fig. 12.11 Semantic image segmentation using U-net (Ronneberger et al., 2015)

masks to train the model and as quality control during validation. The bottom-right panel of Fig. 12.11 shows the predicted mask. Finally, the bottom-left panel compares the predicted mask with the original image to highlight the quality of the results.

Going into more detail on the application of semantic image segmentation to Earth Sciences is beyond the scope of this book. For those interested, I strongly recommend to see the TorchGeo library[3] (Stewart et al., 2021).

---

[3] https://pytorch.org/blog/geospatial-deep-learning-with-torchgeo/.

References

Aggarwal, C. C. (2018). *Neural networks and deep learning.* New York: Springer. https://doi.org/10.1007/978-3-319-94463-0

Badrinarayanan, V., Kendall, A., & Cipolla, R. (2017). SegNet: A deep convolutional encoder-decoder architecture for image segmentation. *IEEE Transactions on Pattern Analysis and Machine Intelligence, 39*(12), 2481–2495. https://doi.org/10.1109/TPAMI.2016.2644615

Cover, T. M. (1965). Geometrical and statistical properties of systems of linear inequalities with applications in pattern recognition. *IEEE Transactions on Electronic Computers, EC-14*(3), 326–334. https://doi.org/10.1109/PGEC.1965.264137

Fischer, A., & Igel, C. (2012). An introduction to restricted Boltzmann machines. In *Progress in Pattern Recognition, Image Analysis, Computer Vision, and Applications.* Lecture Notes in Computer Science (including subseries Lecture Notes in Artificial Intelligence and Lecture Notes in Bioinformatics) (vol. 7441, pp. 14–36). https://doi.org/10.1007/978-3-642-33275-3_2/COVER

Fukushima, K. (1980). Neocognitron: A self-organizing neural network model for a mechanism of pattern recognition unaffected by shift in position. *Biological Cybernetics, 36*(4), 193–202. https://doi.org/10.1007/BF00344251

Goodfellow, I., Bengio, Y., & Courville, A. (2016). *Deep learning* (vol. 29). Cambridge: MIT Press.

Hastie, T., Tibshirani, R., & Friedman, J. (2017). *The elements of statistical learning* (2nd ed.). Berlin: Springer.

Hinton, G. E., Osindero, S., & Teh, Y. W. (2006). A fast learning algorithm for deep belief nets. *Neural Computation, 18*(7), 1527–1554. https://doi.org/10.1162/NECO.2006.18.7.1527

Hochreiter, S., & Schmidhuber, J. (1997). Long short-term memory. *Neural Computation, 9*(8), 1735–1780. https://doi.org/10.1162/NECO.1997.9.8.1735

Imambi, S., Prakash, K. B., & Kanagachidambaresan, G. R. (2021). *Deep leanring with PyTorch.* New York: Manning.

Koller, D., & Friedman, N. (2009). *Probabilistic graphical models.* Cambridge: MIT Press.

Kumar, T. A., & Abraham, A. (2022). *Recurrent neural networks: Concepts and applications.* Boca Raton: CRC Press.

Long, J., Shelhamer, E., & Darrell, T. (2015). Fully convolutional networks for semantic segmentation. In *2015 IEEE Conference on Computer Vision and Pattern Recognition (CVPR)* (pp. 3431–3440). https://doi.org/10.1109/CVPR.2015.7298965

Maggiori, E., Tarabalka, Y., Charpiat, G., & Alliez, P. (2017). Can semantic labeling methods generalize to any city? The inria aerial image labeling benchmark. In *International Geoscience and Remote Sensing Symposium (IGARSS)* (pp. 3226–3229). https://doi.org/10.1109/IGARSS.2017.8127684

Milletari, F., Navab, N., & Ahmadi, S. A. (2016). V-Net: Fully convolutional neural networks for volumetric medical image segmentation. In *Proceedings - 2016 4th International Conference on 3D Vision, 3DV 2016* (pp. 565–571). https://doi.org/10.1109/3DV.2016.79

Ronneberger, O., Fischer, P., & Brox, T. (2015). U-net: Convolutional networks for biomedical image segmentation. In *Medical Image Computing and Computer-Assisted Intervention – MICCAI 2015.* Lecture Notes in Computer Science (including subseries Lecture Notes in Artificial Intelligence and Lecture Notes in Bioinformatics) (vol. 9351, pp. 234–241). https://doi.org/10.1007/978-3-319-24574-4_28/COVER

Stevens, E., Antiga, L., & Viehmann, T. (2020). *Deep leanring with PyTorch.* New York: Manning.

Stewart, A. J., Robinson, C., Corley, I. A., Ortiz, A., Ferres, J. M. L., & Banerjee, A. (2021). TorchGeo: Deep learning with geospatial data. https://doi.org/10.48550/arxiv.2111.08872

Yamashita, R., Nishio, M., Do, R. K. G., & Togashi, K. (2018). Convolutional neural networks: An overview and application in radiology. *Insights Into Imaging, 9*(4), 611–629. https://doi.org/10.1007/S13244-018-0639-9/FIGURES/15

Printed in the United States
by Baker & Taylor Publisher Services